AF886610

AFTERMATH

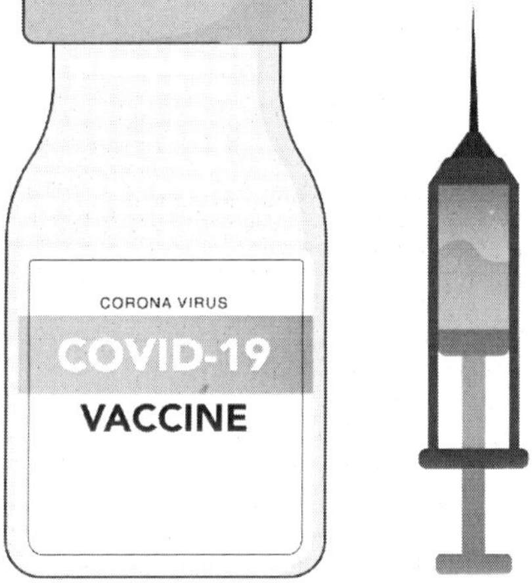

Edited by Scott Tilley, Ph.D.

AFTERMATH

Front cover image from Steven Valentino / Shutterstock.com

Back cover images from NEGOVURA / Shutterstock.com and
CDC / Alissa Eckert and Dan Higgins

Chapter divider from Steven Valentino / Shutterstock.com

Published by the Anthology Alliance

**Anthology
Alliance**

An imprint of Precious Publishing, LLC

Precious Publishing
www.PreciousPublishing.biz

ISBN-13: 978-1-951750-10-7

TABLE OF CONTENTS

Dedication v

Preface vii

Acknowledgements ix

Freedom 1
Anne Bonner

COVID-19 Changed My World 7
Betty Whitaker Jackson

Theurgy 53
Nicholas Kaplan

Somatosensation 63
Elayne Kershaw

Friendship and Love in a Post-COVID-19 World 83
Elayne Kershaw

The Factors, the Problem, and the Product 99
Nanda Klein

Fool Me Twice 127
Mark Lyons

M.E.E.K. 161
Richard A. Marschall

Predictions of Life After COVID-19 183
Ashley McGrath

Rehab and Relocation During COVID-19 189
Richard McNamara

COVID-19 Behind the Scenes in Brevard County.............207
Rick Neale

Learning to Live After the Virus.......................................225
Stacy Strickland

Schools and Businesses Will Adapt to Thrive 257
Dana Thomas Weber

About the Editor..275

Anthology Alliance 277

Dedication

To everyone who:

- *Lost family and friends due to COVID-19;*
- *Suffer from the health effects of long COVID;*
- *Yearn for a post-pandemic sense of normalcy.*

Rest assured – things will get better. They have to.

Andy Dean Photography / Shutterstock.com

PREFACE

What will life be like in the aftermath of the COVID-19 pandemic? Will we reset our "new normal" back to the way it was in the "before times" (2019), or has society fundamentally changed? Will we ever stop using masks? When will social distancing become a thing of the past?

Will we all eventually succumb to the dreaded contagion? Will mass vaccinations and prior infections lead to a modicum of herd immunity? Will the coronavirus continue to mutate with ferocious speed?

Today would have been my father's 90th birthday. He was born in 1932 and passed away in 2018 after a lengthy battle with multiple illnesses. His last years were not entirely pleasant, but at least he did not endure the indignity and isolation of the lockdowns common in long-term care facilities in 2020 and 2021. I don't know what he would have thought about the post-pandemic world of 2022, but as a member of the "Silent Generation," I doubt he would have complained very much. He probably would have focused on living his life as best he could – good advice for everyone in any era.

Scott Tilley

Melbourne, FL

June 29, 2022

ACKNOWLEDGEMENTS

During the height of the COVID-19 pandemic, health care professionals and medical researchers were both praised and derided – sometimes at the same time. They were operating under exceptionally difficult circumstances. Doctors, nurses, and staff were doing their utmost to manage overwhelmed hospitals and long-term care facilities. Epidemiologists, virologists, and scientists worked flat-out to get vaccines developed, shipped out the door, and jabbed into arms in record time. Thank you to both groups of people.

Where would I be without my animal zoo? While working from home for the past few years, my pets have always been at my side. There will be special treats tonight for Grace, Hopper, Ripley, Sherlock, and Watson.

I want to thank everyone who submitted their work to be considered for inclusion in the book; without them, this anthology would not exist. I apologize for the publication being a little tardy, but I think the final product was well worth the effort. As always, any errors remaining in the text are solely my responsibility.

FREEDOM

Anne Bonner

Freedom

Is that what lies ahead?

No longer

Wearing masks

Social distancing

Hunkering down

Virtual learning

Virtual meetings

Behind closed doors deaths spiraling

COVID-19

Suicide

Murder

Human spirit

Wants Desires Needs

Interaction with others

Family

Friends

Neighbors

Restaurants

Theaters

Cruises

Long awaited COVID-19

vaccine purifies the air

atmosphere safe again

Freedom

Elixir of life!

About the Author

Anne Bonner is a fifth-generation Floridian. She has published ten historical fiction books in ten years, all set in the early wilds of Florida. Anne attended schools in Cocoa and the University of Florida. She's a member of the Florida Historical Society, the American Pen Women, and a former Board member of the Space Coast Writers' Guild.

AFTERMATH

Cavan-Images/ Shutterstock.com

Freedom at last! No other piece of "personal protective equipment" (PPE) was more controversial than face masks, also known as "chin diapers." In the aftermath of the COVID-19 pandemic, the use of face masks remains inconsistent. The result is mass confusion.

AFTERMATH

In early February 2020, the New Zealand Government barred entry to most travelers from China in response to the global COVID-19 pandemic that originated Wuhan.

In response to rising cases from overseas travel and within the community, Prime Minister Jacinda Ardern closed the country's borders to non-citizens and non-residents on 19 March 2020.

In August 2021, New Zealand entered nationwide lockdown [again] due to a case of community transmission in Auckland of the Delta variant, with subsequent community cases in Auckland and Wellington. Auckland remained in a form of lockdown until 3 December 2021, when the new COVID-19 Protection Framework came into effect.

https://en.wikipedia.org/wiki/COVID-19_pandemic_in_New_Zealand

Lockdowns are the opposite of freedom.

COVID-19 CHANGED MY WORLD

Betty Whitaker Jackson

Chronology I: Know Your Enemy

My theme for 2020 was "Sharing the Plenty in 2020." Little did I know what would be in store for the world, or why my role as an encourager would be stretched in so many ways. I am beginning this piece on January 21, 2021. A year ago today, the first case of COVID-19 was reported in the United States. The man had been in Wuhan, China. Below are excerpts from the CDC report dated January 21, 2021.

January 21: CDC Confirms First US Coronavirus Case

A Washington state resident becomes the first person in the United States with a confirmed case [1] of the 2019 novel coronavirus, having returned from Wuhan on January 15, thanks to overnight polymerase chain reaction testing. The CDC soon after deploys a team to help with the investigation, including the potential use of contact tracing.

January 21: Chinese Scientist Confirms COVID-19 Human Transmission

At this point [2], the 2019 novel coronavirus has killed four and infected more than 200 in China before Zhong Nanshan, MD, finally confirms it can be transmitted from person to person. However, the WHO is still unsure of the necessity of declaring a public health emergency.

January 23: Wuhan Under Quarantine

In just two days, 13 more people died, and an additional 300 were sickened. China makes the unprecedented move [3] to close off Wuhan and its population of 11 million and place a restricted access protocol on Huanggang, 30 miles to the east, where residents can't leave without special permission. Up to 18 million people are under strict lockdown.

January 31: WHO Issues Global Health Emergency

With a worldwide death toll of more than 200 and an exponential jump to more than 9800 cases, the WHO finally declared a public health emergency [4] for just the sixth time. In addition, human-to-human transmission is quickly spreading and can be found in the USA, Germany, Japan, Vietnam, and Taiwan [5].

COVID-19 teaches us how vulnerable we are. We are literally in survival mode. We question, we doubt, we cope, we despair, and we accommodate. Yet, today's death toll is over 420,000 in the United States alone.

Ultimately, we will experience a changed world in the *aftermath* of COVID-19.

***

In early 2020 we began to hear about the threat. At first, perhaps naively, it now turns out, we thought we could isolate it. Keep the Chinese virus out!

The crux of the problem: the evil has escaped. Hordes of infected Chinese have fled or are dispersed to infect Europeans and Americans.

We hear about New Rochelle, a Jewish community. First, "Put up barricades!" they say. Then, "Close the borders!" they say. "It affects old people with pre-existing conditions!" they say.

Week after week, we listen to Dr. Fauci, Dr. Birx (I admire her scarf collection), Vice President Pence, Governor Cuomo, Dr. Oz, Sanjay Gupta, Dr. Siegal, and other "experts" whose opinions change by the hour. Misinformation leads to distrust; distrust leads to tension; tension leads to worry; worry leads to depression; depression is painful. We abhor crises invading our lives. We want this whole thing to be over.

The "come-ons" for this Call for Submissions listed many aspects of the *aftermath*. Without question, COVID-19 affects doing business, obtaining necessities, practicing religious and cultural freedoms, educating our children, and protecting our elderly. It weighs on our psyches.

Our new President is appointing Dr. Fauci and Dr. Dearlove, among others, to investigate answers to the origins of the virus. But, according to Steve Hilton's report tonight, January 24 [6], these are the very people who commissioned the study and financed a "Gain of Function" virus study in Wuhan in 2014.

Before we suggest urban legends or conspiracy theories, it would do us well to investigate the roles of the National Institute of Allergy and Infectious Disease, and the World Health Organization (WHO), in financing the genetic manipulation of viruses in Wuhan and our universities. Did President Obama and President Trump defund these studies? Just this week, in a speech to the WHO Board of Directors, Dr. Fauci pledged to make up for lost funding from the USA: "We will work with partners to develop new international financing mechanisms for health security," which is government-speak for continuing genetic and biological studies of new viruses produced in laboratories [7].

Now, in January 2021, we hear about mutations to the original virus, some more contagious, circulating in places like London, Germany, South Africa, and Brazil. As a result, travel bans have again been issued.

Perhaps more frightening, and therefore postponing the *aftermath* we seek, is that treatment against the mutation may be ineffective. The CDC reports: "Decreased susceptibility to therapeutic agents such as monoclonal antibodies" in mutated forms of the coronavirus [8].

An *aftermath* is that we question and doubt information. We consider or reject "urban legends" or disguised misinformation. Our emotions overreach our wisdom. We are skeptical, critical, antagonistic, and suspicious because of COVID-19. In a mere year, we've lost our individual freedom: we wear masks; we isolate; we gather in small groups; we self-quarantine; we huddle in fear; if we're Republicans, we dare not speak.

Change is difficult for people. Mandated quarantine is anathema. Palpable fear invades our lives, especially for seniors like my husband and me. Suspicion of others spreads to canceled activities and altered lifestyles.

Parents leave jobs to attend to their children at home, a real gamechanger for family finances, routines, and patience levels.

As cases increase among us, we dread reports of daily tolls. How many are tested? Are the tests accurate? Why does it take days to get results? Why are there results for people who never took the test? Are all the deaths actually COVID-19 related? Have 420,000 Americans died this year of this one cause?

***

Our grown children teach music in Brevard. Locally, by March, students cannot return to school following Spring Break. Teachers train to teach remotely. Protocols for home settings (no views of bedrooms), class participation by students, and submission/return of work procedures are established.

Buildings are prepared and inspected according to CDC standards. Deep cleaning systems, air purifiers, masks, sanitizers, and personal protective equipment (PPE) are obtained through federal emergency grants. Without much leadership, states are thrust into battle without much input from the generals.

Beleaguered states' officials leave decisions up to localities. That includes the school board members who hotly debate whether masks are recommended or mandated. Parents make difficult decisions regarding the district's homeschooling, remote, or in-person instruction. Private schools choose their options.

Depending on the strength of teachers' unions and community outrage, some schools in our state open; some stay closed. Some, like ours, try to accommodate everyone to the distrust of almost everyone.

One *aftermath* is that teachers, with their fantastic flexibility, creatively teach both in-person and remotely, knowing they will be exposed to this marching enemy every day.

Hospitals cope, calling in staff, quickly training them in protocols, and wearing the new protective coverings – if they can get them. Administrators are overwhelmed and are forced to schedule twelve-hour shifts because they are short-handed. Patients, critically ill and in dire straits, have tried to self-medicate. Few understand the immune system is attacked by this thing, that inflammation and pre-existing conditions exacerbate attempts to save them unless they are treated early.

A positive *aftermath* is that we respond well to the need. American ingenuity again results in problems being solved. Ventilators? Whole industries tool up to create life-saving equipment on short notice. Warp speed. Can a vaccine be developed? The usual timeline is years. The *aftermath* is immense creative thought thrown at the enemy. It's a beginning of a call to arms.

War against the virus imperatives: Shut down the economy. Wash hands. Keep socially distant. Mask. Isolate. Self-quarantine. Seniors become hermits, hiding in our bivouacs.

In Florida, we shop before a hurricane. It's just what we do. The result is the same: empty shelves! As in any crisis, self-preservation hoarding takes hold. Food and paper product prices rise dramatically. Toilet paper is nowhere to be found.

***

Into this thing for a year now, we know families who have lost income, businesses, and ultimately, hope. We know those whose loved ones have died alone. Loved ones mourn and grieve with survivor's guilt. There are no goodbyes and no funerals, just empty chairs, lost dreams, and no closure. This *aftermath* affects generations.

We know those so sick, they wished they could die.

Daily we hear of families and friends with positive tests. How many more pots of soup will I deliver to the sick? How many more meals will I leave outside the door, longing to hug the ones I love?

Some official somewhere asks, who in our society is genuinely "essential?" What a blow to the ego! We have a whole new measure of our success and worth in this *aftermath*. Stay home.

Engineers, we know many of them, think years of learning sophisticated techniques and ways of thinking, working collaboratively, is the essence of "essential." After all, teachers have taught cooperative learning, group work, "getting along" skills, and brainstorming to solve problems. But suddenly, degrees and high salaries are not the measures of "essential."

The engineer works from home, often while teaching/babysitting, trying to accomplish their little piece of the work contract. The security-clearance safe room is miles away. So now he knows only enough of the sequence to meet their frustrated boss' daily quota.

Then that supervisor is also sent home to work.

How's that going? Are your kids learning remotely? Have you figured out how to teach your fourth grader the new way to attack division problems with all your engineering degrees and expertise? Do you put all the little boxes in the right places? Better yet, you who can't read cursive, try teaching it! Then you hear, "That's not how my teacher said to do that," and tears, stamping of feet, and door-slamming. The *aftermath* becomes heart-wrenching and very real when our little ones learn to adjust to their changing, confusing, frightening world. Worse yet, is dad or mom, previously the source of all knowledge and truth, diminished in that child's eyes?

Important question: Will appreciation for teachers be a positive *aftermath*?

The "essential" person stocks shelves (if there's toilet paper available) or wipes the handles on shopping carts for hours at a time. The "essential" worker checks out groceries behind plexiglass and wipes the card swiper if they remember to do that. Note: a year later, that "essential" worker can't get the vaccine either.

Will the new normal be normal?

***

Our son, skilled in running the sound/vision system at church, helps create live streaming worship services. We're thankful technology helps, but it raises *aftermath* questions. Will parishioners worship together again? Will they "watch" and have brunch in their "jammies"? Will they ever return to communal worship?

In late May, we offer in-person worship for those daring enough to attend with all precautions in place. Our little children, usually excused for small group activities and study, adapt to working on packets of material near their families instead. Families with any hint of infection are urged to stay home. Next week we put in place a completely masked section for those still afraid to return. Have we seen positive cases? Yes. Some among the staff.

As people we know get infected, and contact tracing includes more and more of us in webs of uncertainty, the *aftermath* of a year's struggle with this beast is beginning to fray nerves and cause discouragement.

What about the survivors? We know those with chronic and frightening symptoms after having COVID-19. Many become depressed, even suicidal. Near-death experiences show us how very vulnerable we are. We know that within days, each breath could be our last. The cloud hovers over our minds. Who will be next?

We all know medical personnel who have battled against this relentless thing long and hard. Each one has feared for their family. Many self-isolate. What is the damage to a child who no longer hugs their mom at night? How do marriages survive? Each one is concerned about physical and mental health. Dealing with daily deaths and complicated recoveries in patients is exhausting. It takes a brutal toll. We pray for doctors, nurses, technicians, administrators, CNAs, and cleaning crews every day. We can't imagine how EMTs and funeral directors cope. They have not even reached the *aftermath* yet.

I'm trying to get vaccinated. I've spent hours in phone jail – on hold, listening to terrible music, before dawn, with two-hour second-by-second countdown clocks. Then: no appointments available.

Finally, after a contentious election, adding insults to injury, a few instances of perceived misuse of policing power, continuous frustration, and pent-up stress lead to unprecedented anger, unrest, separation, ostracizing of friends and families, and riots in our cities.

A significant result, an *aftermath* to this season is a sense of seething anger, even anarchy, waiting just beneath the surface for an excuse to break forth in civil unrest. Emotions are hidden, yet so very real, smoldering just below the conscious awareness. It's like constant fight or flight syndrome. Under our masks, we've forgotten how to smile. We no longer converse. We lose our identity. That is threatening.

***

Perhaps the most alarming *aftermath*, the threat I fear most except falling prey to this disease – me or my family – is my loss of freedom.

An *aftermath* of our year of COVID-19 is an election that deeply divided the country. I admit that I am on the "losing" side of the outcome. However, I am not a cult follower of Donald Trump.

Those now in power want to vilify, punish, malign, denigrate, and even silence those who supported Donald Trump's conservative program. A result of January 6 is a rushed judgment of impeachment, removal of the President from social media expression and voice, and a perceived threat against any who opposed President Biden's election. His calls for unity while systematically enraging many with his forty-plus executive orders dismissing programs and policies without legislative debate further divide Americans.

Sure, the *aftermaths* of COVID-19 will show their influence affected all aspects of one's security: business, education, economy, family dynamics, mental and physical health, safety, and religious practice – all pillars of society. It also deeply divided the country, ended a contentious Presidency, and caused worldwide misery.

History will write the *aftermath.* Will man's

creativity, innovation, compassion, reasonable discourse, cooperation, and decision-making solve the tribulations a minute virus precipitated?

Chronology 2: COVID-19 Affects My Family

By God's grace, my husband and I are self-quarantined octogenarians. We visited our children and grandchildren in Rockledge from January to March by Skype. Schools were closed. Our music-teacher son and daughter-in-love struggled with at-home teaching while supervising at-home learning for their fourth- and first-grade children. Our other daughter set up her virtual classroom on our back porch.

By Easter, the reality of COVID-19's threat was becoming very real; normal was anything but. We saw our grandchildren on Sundays, worshiped together remotely in our home, had lunch together, and prayed for our protection. We feared infection for any of the seven of us.

During isolation, hubby and I completed our "honey-do" list around the yard and house. We shopped early hours at Publix, sewed 30+ masks, attended the church's live stream, and taught our church covenant group, most in their seventies and all with chronic "pre-existing" conditions, to Zoom.

By keeping in touch with these three widows, one widower, the couple caring for parents in their 90s, and shepherding four of our group through hospital stays during COVID-19, we have maintained close contact, virtually, through Zoom and email. We even have members in Ohio and Texas. What would we have done without technology throughout this pandemic?

***

We watch too much television drivel, read all the books stacked in various places, and unfortunately, send more sympathy cards than birthday greetings.

My sister and her husband visited us in April, even though we urged them not to come because of Brevard County's infection invasion.

They have an RV; they insist they are self-contained and aren't going to wear masks or take precautions. They feel safe in their confined world, their cocoon. They are not concerned about COVID-19. We, however, are and maintain our "healthy" precautions. Their statement, "We're covered by the blood of the Lamb," says it all. They, therefore, feel they are impervious to the virus' attack.

Then, they fly to Ohio to help move their daughter to Lakeland. Full plane. Crowded airports at both ends.

They borrow masks from us. I'm not sure they wear them. Then they drive Deanna and her belongings by U-Haul, stopping along the way, then return here to Wickham Park to their RV a week later.

My husband and daughter stayed their distance from Ellen and David. I, however, worked closely with Ellen. Her "thing" is genealogy, so we continued our family history studies. I compiled our "family tree" legacy back to the 1066 Battle of Hastings, and we became fascinated with our mother's manuscripts. I had postponed reading them since her death in 1997. Quarantine presented the perfect opportunity. Ellen and I were amazed at what we found.

A week after Ellen and David leave, I get a rather frantic phone call. They are in a small town in Georgia when they both test positive for COVID-19. David lands in a Podunk-center hospital where he is in their ICU. He might have died if he had not been for a new treatment, Remdesivir, and an antibody infusion.

Ellen, who has never driven the RV and hadn't handled a stick-shift car since high school, is ten miles away from David's hospital bed. She tests positive for COVID-19. Their campsite is miles away from stores. She maneuvers there, gets lost several times but stocks her cupboards for what she thinks might be a day or two. She is not able to visit the hospital. She receives no treatment. It is a scary time. Deanna, in Lakeland, tests positive too.

After six days, David is released from the hospital to bed rest. He must return to the hospital for nebulizer treatments, x-rays, and observation. They admit that there is no room for him to stay, although he should remain hospitalized. Ellen becomes both patient and nurse. She learns to administer oxygen, numerous meds, and COVID-19 becomes very real. They are quarantined for two weeks more. They are too tired to travel anyway.

Meanwhile, we are counting days on our fingers. Are we, too, infected? As days pass, we are encouraged.

***

In some ways, life has become simpler, concentrating on the necessities. For old folks like us, that's a welcome thing. But withdrawal from life steals our joy, purpose, and incentives to achieve. Our calendar is blank. Sometimes we don't know what day it is. Why wash the kitchen floor? No one is coming over.

Too many "stay in jammies" days. No haircut since last March. Why wear makeup? No one rings the doorbell. Even the Amazon guy sends us a message telling us he's delivered a package. What happened to ringing the doorbell or stopping for a chat? We stay glued to our phones out of monotony. Not that we talk to anyone.

My routine changes. I usually meet with a ladies' group at Cracker Barrel on Tuesdays. A few of them still meet. They're all younger than me. If I get COVID-19, my teacher-family will be exposed. There are no substitute teachers to take over their music classes; it's too risky.

Our favorite café, our go-to takeout place, has closed twice in the last six weeks. A positive case. Then another. Yet another. The owner has time to put in new flooring. That's about the only positive thing we've heard. We miss our Friday dates at his little place. We'd considered it "safe."

Then comes November. There are no concerts to volunteer for this year, although the Symphony does have drive-in concerts we attend. There is no Christmas music to learn for the choir. There are no Christmas parties, no school concerts, and no soccer or basketball games. Although we support some of our favorite places ordering takeout, we dared not go into restaurants.

There is no mother-daughter dress shopping for the perfect holiday outfit. In fact, sight unseen, we shop online. As a result, we see more of the Amazon Prime delivery guy than our friends.

Our only "out of the ordinary" is a big negative. We are victims of a scam, hack, or fraud against our bank account. We spend weeks changing all our account numbers, especially our online payments. What a hassle! That at least gives my husband something to do while I write a book!

On a personal note, while we've been restricting our excursions away from home to what we consider "safe places," we've carried on our lives in isolation. However, my husband's duties as an elder of our church have kept him extremely active, if sitting in his desk chair and Zooming can be considered active.

***

Through countless meetings, a great deal of prayer, and God's guidance, the elders have negotiated an agreement to sell our church building and grounds, including the Christian school we've sponsored for almost thirty years, to Pineapple Cove, a school in Palm Bay. In addition, we will build new facilities at the other end of our property. This new vision will serve as an *aftermath* of God's leading for the future.

We've also sponsored four new church plants during

the pandemic. In addition, our seminary is preparing to launch another new pastor, a recent seminary graduate from our congregation, to begin our second local, new church plant.

Our missionary conference this year highlights Zoom meetings from across the world. Unfortunately, we cannot invite the twenty or so church leaders from abroad to attend our usual congregational meeting. The meeting is usually a big deal, featuring foods from all the areas from which our guests come: Indonesia, India, China, Brazil, Argentina, France, England, Chile, Miramar, Ukraine, and others. Most feature indigenous pastors, and several of these new church plants have already begun church plants as part of their mission.

Our website says: "We are committed to planting churches globally and locally. By our church's 50th anniversary in 2028, we have a vision to help plant 50 new churches worldwide, with at least 3-5 of those churches being planted in our own backyard." So, we're well on our way!

We have partnered to establish three new congregations in former Soviet Union countries during this pandemic. Even when some of our families have suffered financial losses during the pandemic, our Faith Promise giving to missions and other outreaches continues to flourish. We consider this an *aftermath* in our ministry history.

In December, our congregation contributes over 250 Operation Christmas Child boxes to Samaritan's Purse and provides 110 gallons of frozen soup to His Place ministries in Melbourne to feed people. Although their cold night shelter has been closed this year, their clients are still fed. In addition, our members have adopted four local schools, helping with landscaping, painting, and providing supplies, gifts, teacher appreciation events, and meals. So now, when sharing is essential, we support local and distant ministries, broadcast live streams, and accomplish our ministry goals. Even in times of crisis, and perhaps as a result, these are positive *aftermaths* to which we will keep contributing.

The pandemic has opened our eyes to challenges. Our Stephen Ministry is busy counseling people whose spirits are affected by isolation: those adjusting to difficulties living together 24/7 in families with discord, those who have lost employment and businesses, and those who grieve. We counsel and supervise programs for those with addiction, those overcoming or dealing with sexual abuse, and our isolated widows and singles living alone. Our mercy ministry reaches out to our own members and others in the community.

In the *aftermath* of COVID-19, we've adapted. We've found a cooperative spirit in accomplishing complex tasks. We've created. We've accommodated. These have strengthened us and will continue in the *aftermath.*

***

Our two grown children and our daughter-in-love are orchestra musicians and teachers. To say local musical groups have been suffering through this time is an understatement. For some musicians and artists, contact with an admiring public means they get paid. Not so much now. Their income has suffered.

Teaching students to play the violin or cello remotely is difficult. I admire the way teachers have learned new skills to teach Brevard County students remotely and teach others in-person in modified classrooms at the same time. It is immensely stressful.

They've become part of the school's custodial staff, wiping down desks, instruments, and chairs between classes. They must keep students from pre-kindergarten through twelfth grade distant from each other. The teachers must wear masks, which are restricting but also force them to teach in new ways to encourage the students to make and enjoy music. It's an enormous task.

Orff instruments are silenced. Orchestra concerts, which inherently present motivating experiences for students to practice, have been canceled. All State, a supreme honor for achieving students, is virtual. So is the music conference for teachers. They're still proposing a one-day All County program for auditioned students. They're still seeking approval.

Choruses can't sing. Students have a new understanding of playing in ensembles when separated. Yet, our teachers try to stay motivated and enthusiastic while watching their star musicians languish. Their self-worth morale is at a new low. They are tired, concerned for each other and their students, and frustrated.

In our circle of friends, we know of a suicide attempt after a COVID-19 survivor was prescribed an anti-depressant! A mother and daughter committed suicide a day apart, unable to cope. My daughter played cello at the funeral. She had taught this student, a recent graduate, for eight years. She also lost a recently retired colleague; another young teacher is on a ventilator in an induced coma. Many of her students were sent home to quarantine. That affects preparation for solo and ensemble festivals for upcoming judging and filming orchestras for Music Performance Assessment.

We've had to babysit and supervise at-home

learning for our granddaughter. Someone in her class was exposed, not ill, so half the class was sent home. Teachers have home packs ready. I called them "busy work" in my forty-year teaching career. Detest that!

Kaley does get to attend her mom's music class, but virtually. They can't sing; they can't move around the room; they practice clapping out rhythm patterns. They watch a film about a composer. Ten students are present; ten are at home, wishing they could be with their friends. I monitor temperature, hoping there's no sign of contagion in my sweet nine-year-old. At least she doesn't have to wear a mask all day. I take my chances.

Our daughter changes her teaching outfits before coming to our house for dinner each night. She's been exposed to music students in eight schools in Palm Bay. She could, one day, be a super spreader! She worries about that. It's scary. Every day we get a report about whole sports teams sent home, or e-learners who have caught the virus somewhere in the community. They would have been safer in school, she contends.

***

Our family and volunteers have nurtured the Space Coast Symphony Orchestra for twelve years. January 16-17 was their first indoor concert weekend since COVID-19 hit, with only twenty-five percent audience allowed. During the summer, they performed outdoors, both in Viera and Vero Beach, with people sitting in their cars tuned to a radio station or spread through parking lots, socially distant. Additionally, performers recorded themselves on YouTube for audiences to enjoy their performances and hopefully elicit contributions to keep the orchestra afloat. Their next performances are tonight at Satellite High School and tomorrow at Vero Beach High School.

We won a sizable grant for children's music. Implementation has taken great innovation, but we've done it! As a result, our children's programs have continued, but remotely. For example, the orchestra members spent a day recording music for our annual children's outreach program at great peril to their health. However, every fifth grader in Brevard and Indian River Counties heard Michelle Campanelli's illustrated story "Music at the Watering Hole," recorded by the Symphony and presented by film to the schools as part of the grant featuring children's music.

It was not quite the same "live" performance they've attended in previous years. Still, we innovatively accomplished the purpose through ingenuity and great personal sacrifice by the musicians and technical staff. This year, the film has been shared with all students to augment elementary and secondary music class resources. So, we have reached all the district's students without the original intention. That's a first! The pandemic has taught us to be flexible, creative, and determined to overcome obstacles. Perhaps that is what we were supposed to learn from a time such as this. We pray that, as an *aftermath*, orchestras and other creative arts survive this pandemic and its restrictions.

When feasible, quartets of symphony players have introduced children to instruments in several locales, including the Brevard Zoo and McKee Botanical Gardens in Vero Beach. But, again, these are outdoors, and children and musicians are safe during this time.

Director Aaron Collins has shepherded the orchestra, determined to stay on the mission through this crisis. Several performance spaces have been closed, so adaptation and struggles for venues continue, and there are many hungry musicians eager to perform together. We can only hope the *aftermath* will bring enthusiastic audiences to support this group once more. We have dared to schedule monthly indoor concerts for each upcoming month, even during this pandemic. We're courageous, even daring.

***

Aftermath? Schooling will continue, yet nothing will ever be the same. Musicians, artists, restauranteurs, and shopkeepers will struggle as they always have. Tempers will cool, and people will ignore the out-of-touch politicians in Washington until we get riled up again in four years, and the "loyal" opposition will shudder as dictates require.

But writers, like me, and books like this one, will attempt to make sense of the world. And may we all unite to applaud Amanda Gordon's poem recited with such passion on Inauguration Day.

We will one day hug again, maybe even shake hands. We will smile again, not just behind masks.

But we shall never forget, just as events in the past have scraped man's consciousness raw: wars, assassinations, 9-ll, Pulse, January 6, an inauguration in a militarized zone, and COVID-19.

It's just that the last one permeates our lives and perhaps takes them.

Chronology 3: COVID-19 Affects My Writing

I am mildly excited when a fellow writer enthusiastically announces NaNoWriMo in the fall of 2019. We are attending our monthly meeting of the Space Coast Writers' Guild.

When this speaker finishes, I know November is called National Novel Writing Month. The acronym NaNoWriMo encourages writers worldwide to write 1500+ words a day, completing a novel in a month.

The website offers incentives: book outline suggestions, writing instructions, buddy groups, and graphs and charts to measure the writer's daily progress. They even award badges for reaching goals.

***

I just completed my twenty-third book, *Apple Tree Lane.* I keep thinking about the characters; I love them.

The setting is the community where I grew up, Columbia County, New York, on the Hudson River. For *Apple Tree Lane,* I research the life of the Mohicans who first occupied the land and read some of the tales of the Dutch and English settlers. I continue to learn many interesting things, such as why there are so many lilac bushes and how the indigenous Native Americans use herbs and plants for medicinal purposes.

I reread Washington Irving's "Rip Van Winkle" and the local stories about Spook Rock, and research hundreds of place names from the indigenous tribes in the region [9]. I know Poughkeepsie, meaning "the reed-covered lodge by the little-water place" [10]. And

Coxsackie (meaning owl hoot) once taught at Ockawamick Central School.

I remember tales of my father driving his first car across the Hudson River ice from Athens to Hudson one winter. That was long after ice harvesting occurred until electricity arrived in the 1920s and 1930s and literally dried up that business.

But these sixty acres on the Hudson River once housed one of the river's 125 ice houses. I know the indigenous Mahicannituck or Muhhekunnetuk people were called the people where "the river that flows two ways," or "the water that is never still" lived. The Hudson River is an estuary, changing with the tides, including both salt and fresh water [11].

One October morning, I woke up from dreaming about developing these sixty acres on my imaginary Apple Tree Lane into a place that respects its history and preserves its environment for others to enjoy. I can't get the story out of my head.

Serendipity is when two things occur simultaneously, usually without planning but quite by surprise. At that SCWG meeting, the proverbial "lightbulb" flashed, and I knew I wanted to write a sequel called *At Home on Apple Tree Lane.*

If I can clear my calendar, I am determined to do the NaNoWriMo thing as a challenge.

***

I am mostly ready for the holidays. Our son and daughter-in-law in Rockledge host Thanksgiving. We bring the sides. Easy-peasy. Space Coast Symphony, my place to volunteer, fits into the schedule with the *Messiah* singalong. Preparing Christmas music in the church choir was only a minor interference, and I am sure I can maneuver through a packed season of parties, celebrations, and gift-giving. It is our 51st anniversary, not our 50th, so all that commotion is reduced to a dinner out with the family. That occurs anyway, busy or not. It's just part of living where family is.

Why shouldn't I devote November to writing a sequel?

To say the book just "flows" is an understatement. I love researching Hendrick Hudson, the original Livingston/Van Rensselaer plat of land, and the Van Eyck family, who built the little log cabin in the 1800s and ran the ice harvesting business there. I vicariously help Beth write six books and escort her (and my) *Whispering Woods* four-book series through filming for the world to see [12][13][14][15].

Again, she takes on my persona, or rather I invade hers. Her books appear on the New York Times Best Sellers list, and her middle school books are top American Library Association prize winners. She is earning thousands in royalties; I am not.

I jealously wonder how great it would be to finance the development of property through Jim's successful textbook sales, his honoraria, his professorship, his piano CDs, and her sizable royalty checks, movies, contract advances, blogs, and podcasts. They're living the dream! She has thousands of followers, a fantastic publicist, an eager publisher, and loyal advertisers. It's fiction, right?

In this second book, *At Home on Apple Tree Lane,* I include building the Walters' 2,200 square foot dream home, honoring the shape of a Mohican wigwam with five longhouse-style wings radiating out from the central wigwam-domed Gathering Room. I refurbish the stone-walled Icehouse into a concert and event venue. I include tracing Mohican customs, producing maple syrup, cider, and honey, offering community gardening, encouraging charitable giving, maintaining the orchard, performing classical music, and highlighting the multi-talented families I create. It becomes a thriving business. I love every minute of creating the story.

With this book finished, revised, and published in December, I completed four books in 2019, two of them in the setting of Apple Tree Lane.

***

In March, I am to participate in a celebration of World Poetry Day. I published a book of poetry called *The Nature of Things* [16], and eagerly await participating in the panel discussion that will feature fellow poets in Brevard County.

Alas, COVID-19 interferes. The event is canceled.

After my sister's visit in April, I started reading my mother's manuscripts. She was raised by her sainted grandmother in Texas, moved to Hudson, New York, married my father, and left behind the four of us and now a genealogical legacy of twenty-five. From her second marriage, after our dad's death, she added three adult stepchildren and several step-grandchildren.

We find remarkable stories of her volunteering, are entertained by her children's stories, and learn about Richard's and her obsession with competition kite-flying adventures. I am brought to tears hearing about her parents' rejection of her and several encounters of sexual abuse. We never knew.

"There's a whole book here," my sister says, and indeed, that's how my May to October study and writing *Her Name Was Bess* [17] occurs. It is a combination memoir and tribute to her "I Can Do It" attitude of overcoming dire circumstances to accomplishing outstanding objectives once she puts her mind to them. I include several of her manuscripts.

Thanks to Scott Tilley, another diversion presents itself with his *PANDEMIC* anthology [18]. Our church children's director devises a church-wide competition in March called Isolation Olympics. Tasks are designed as family activities to accomplish in two weeks, while our kids cannot return to in-person school after spring break. It keeps boredom at bay.

The interesting thing is that the old folks like me win several prizes. Our ingenuity pays off. When I say "win," the reward is toilet paper!

I recount the tasks in a poem that is a tiebreaker for fourth place. We must do a project using five assigned words or sayings in an unusual way.

The prize winner is a reenactment of a college piano-saxophone duet – another family skit. This diversion becomes another opportunity for writing and publishing my twelfth anthology piece.

I pretty much spend 2020 trying to be an encourager by posting scripture and positive blogs on my website and Facebook, writing another book, gardening, learning the new technologies of Zoom, Skype, and Roku, livestreaming, discovering RightNow ministries, and becoming thankful for these new ways of communicating while being confined to our 1,600 square foot home in the scrub forest of Palm Bay.

Then comes November. NaNoWriMo! 2020.

Another November, another sequel. This one is *Cortland Place on Apple Tree Lane.* My books, *Apple Tree Lane* [19], *At Home on Apple Tree Lane* [20], and *Cortland Place on Apple Tree Lane* [21], are basically chronological.

Now to the *aftermath.*

I'm debating whether to add the conflict of dealing with COVID-19 to the story, imagining with dread the limitations on all this family has built. How will COVID-19 affect their business, their hopes and dreams, and will financial disaster occur if the story extends into the 2020 environment?

My Pollyanna vision is threatened by the family I've grown to love in the place I cherish. What will happen to the therapy program if children with handicapping conditions cannot come to ride horses or interact with bunnies? What will happen when nursing homes and senior residences are closed to Papa Skip and Dad Joe, who can't take their docile dogs and cats to comfort elderly, lonely residents?

For whom will Kathleen provide gourmet meals in the high-end restaurant, Apple Core? Who will attend concerts, weddings, fund-raisers, and square dances in the Icehouse? Can people work out in the pool, the track, or the exercise rooms in "Time Out?"

Can crowds of people gather for fireworks and barbecues for Memorial Day and the 4th of July? Will Jim's interns be able to build the Mohican replica longhouse and wigwam this summer or discover artifacts in the midden dig across the street? Can the riverboat dinner cruises to the Lighthouse continue? Will Jim perform Chopin in Carnegie Hall in December? I think not!

The *aftermath* in my book will be very real. Nearly as real as "real life" – is that possible? Are my vicarious adventures in my almost completed book my "discover what I know" reason for writing it in the first place?

My first thought is to complete the book by

Christmas 2019. But this Call for Submissions, *Aftermath*, has initiated my thinking about how the Walters family might, could, would deal with the pandemic. So, do I dare include it?

So, how to keep encouraged is the big challenge. Every year I have an adage I create to keep me focused on my writing, my devotions, and my life: "Living the Dream in 2019," "Sharing the Plenty in 2020," and "Celebrate '21, Everyone! God's Got This!" for this year. My blogs, Facebook posts, and stories often emphasize the themes. We learn new ways to do things. We adapt. We humans are amazingly creative and inventive when conditions call for change.

I am naturally an optimist, knowing God is sovereign, and although we can't see the "Big Picture" or answer the "Why?" questions, we can do our best to live life fully and purposefully.

***

COVID-19 has been a unique challenge because we are curious people. We want to know what caused it. Was it a biological weapon to intentionally create worldwide challenges? Why couldn't we stop it when it first appeared? How could it separate people from each other and cause such panic? Will we ever get back to normal? What do we say to those who ignore the warnings? Are we too suspicious of others? Why can't the news broadcasts report other news rather than grim statistics? The biggest: will we ever be free of this, or will there be another on its heels? Can the world economy sustain this threat and others?

The purpose of this *Aftermath* volume is to list so many possible outcomes.

Thinking people want to know. Political upheaval, Deep State rumors on both sides of the political spectrum, distrust, and misinformation tease our consciences. Billions of dollars squandered on political campaigns could have helped millions out of work and thousands of failing businesses. Instead, we've been bombarded with nastiness, innuendo, sarcasm, bias, and words taken out of context—by both sides. Ultimately, we elected a candidate who wisely stayed out of the fray in a Delaware basement.

People learn from their own realm of experience. "So and so had COVID-19, and he's fine now." "It's no big deal." "I'm just going to live life like usual." Or those who live in denial until they become deathly ill. Others are discovering troubling residual effects months after the initial infection. They now have, to their chagrin, a pre-existing condition. Some are feeling ill after their second dose of the vaccine. Some have died! These describe the *aftermath* as well.

COVID-19 is very real. It's an underlying sonorous drone like a bagpipe, always present. We hear it when we wake, the first thing on the news. We spend half the day trying to get an appointment for the saving, we hope, vaccine. Then we hear offensive statements from the anti-vaccine crowd. COVID-19 is in our bedtime prayers each night. We worry when we cough. Allergy? Cold? COVID-19?

I am writing this on January 21, 2021, the day after the inauguration when the Washington mall lit luminaries to represent the deaths our nation has reported, more in one year than in all of World War II. It is sobering. How many have been buried in this world, even in unmarked graves? How many people, how many families in our schools, cities, states, and countries weep and grieve? How can we count the cost of those losses: lost dreams, lost talents and abilities, lost hope, lost potential in those lives? How many books remain unwritten, how many careers ended without final breakthroughs, how many never reached retirement, how many life partners are now alone? We cannot calculate the true human loss in the *aftermath.*

Writers like me have seen an assault on free speech, severe censorship, and offenses by government and social media outlets promising retribution. Our posts are challenged. Our accounts are closed. As in previous societies, will suppression lead to uprisings and chaos?

Do I dare express my outrage, even here?

***

Facebook, Twitter, and other social media services have squelched speech. Like some of my friends, do I boycott these (those that are still available) to express my displeasure? No, for then, I lose my voice altogether. Do I boycott Kohl's or Bed, Bath, and Beyond because they refuse to sell Mike Lindell's pillow product? No, because I already self-boycott Target for its announced abortion and gender-neutral stance. I still have a choice; I still have a voice – although somewhat muffled.

Editors choose which "letters to the editor" see the light of day. I concede that they always have, but only one point of view now seems to get printed.

We have no Wittenberg door where we can post our 95 theses nailed for widespread view.

TV's talking heads tell us what to think, what to believe, and in my case, because I dare to represent a contrary point of view, they subject me to ridicule and intimidation. They call me idiotic (I'll match my degrees to theirs) and angrily shout their denunciations until I exercise my authority to shut them off. My ire remains. I contend that elections have consequences, and COVID-19 affected this one. My personal *aftermath* is painful.

I, for one, reject this aspect of the new normal and suspect I may have lived too long, when people's voices are hushed or disciplined by some business or agency or government enforcing a single point of view. In some "enlightened" person's mind, I may be one of the "deplorables." I may, in Katie Couric's opinion, need "reprogramming." I am repulsed by the name-calling and categorizing.

***

As a citizen, I am appalled that our capital has become a militarized zone. This is part of my post from my blog today on www.bettyjackson.net:

"These days, tempering my temper, few things get me riled, but yesterday's treatment of the National Guard troops in Washington, DC made me seethe. Beneath it all is my dismay at seeing our nation's capital seem like a militarized zone. But to take our husbands, fathers, and sons away from their businesses, studies, and homes, force them to sleep crowded together during COVID-19, and dismiss them to a parking garage in 38- degree weather without adequate facilities makes me angry."

There are frightening statistics about COVID-19 infections among the National Guard troops in today's noontime report. When and if they are dispensed back to their home states, will they carry COVID-19 to their families and communities because of the conditions they were forced to serve in (and live in) Washington? Now there's an *aftermath!*

COVID-19 has stolen my peace of mind, along with my individuality. I am now part of group-speak, the antithesis of my pursuit of happiness and free speech as a citizen of the United States of America.

Furthermore, I will NOT end a prayer, "Amen and A-woman," Emanuel Cleaver, even if the House of Representatives holds this as a new standard of language acceptability. No Sir!

I will pray for both Presidents. I probably will complete *Concord Place on Apple Tree Lane* with coping mechanisms for dealing with COVID-19. No matter what, I will try to be an encourager and continue to blog and post with passion until the powers that be attempt to silence me. Then, I will dare speak again with others in this society who need to find their voice, with all the ingenuity that remains, and all options seen and unseen.

And I can still write this piece for *Aftermath*, with my thanks for allowing my essay to see itself in print.

References

[1] https://www.cdc.gov/media/releases/2020/p0121-novel-coronavirus-travel-case.html

[2] https://www.ajmc.com/newsroom/what-were-reading-trump-frustrated-with-hhs-secretary-humans-can-transmit-coronavirus-premature-death-linked-to-coal-plants

[3] https://www.ajmc.com/newsroom/what-were-reading-insurers-plan-to-manufacture-drugs-trump-hints-at-cutting-medicare-china-quarantines-wuhan

[4] https://www.ajmc.com/newsroom/what-were-reading-global-health-emergency-declared-jj-pay-in-pelvic-mesh-suit-ban-on-fentanyl-lookalikes

[5] https://www.ajmc.com/about/ajmc/journal/about. Accessed January 21, 2021, 10:00 a.m.

[6] https://video.foxnews.com/v/6225847837001. Viewed January 25, 2021, 9:30 a.m.

[7] https://www.hhs.gov/about/news/2021/01/21/dr-anthony-s-fauci-remarks-world-health-organization-executive-board-meeting.html. Accessed January 25, 2021, 9:45 a.m.

[8] https://www.cdc.gov/coronavirus/2019-ncov/more/scientific-brief-emerging-variant.html. Accessed January 25, 2021, 9:30 a.m.

[9] Ruttenber, E.M. *Footprints of the Red Men*, 1906.

[10] S. Buff. *Insiders' Guide to Hudson River Valley*. Globe Pequot Press, 2009.

[11] https://www.mitpressjournals.org/doi/pdf/10.1162/tneq_a_00845. Accessed January 29, 2021, 2:30 p.m.

[12] Jackson, B. *Whispering Woods*. LifeRich Publishing, 2015.

[13] Jackson, B. *Winter Wonderland at Whispering Woods*. CreateSpace, 2015.

[14] Jackson, B. *Splendid Springtime at Whispering Woods*. CreateSpace, 2016.

[15] Jackson, B. *Christmas is Coming at Whispering Woods*. CreateSpace, 2016.

[16] Jackson, B. *The Nature of Things: World Poetry Day 2020*. Independent, 2020.

[17] Jackson, B. *Her Name Was Bess*. Independent, 2020.

[18] Tilley, S. *PANDEMIC*. Anthology Alliance, 2020.

[19] Jackson, B. *Apple Tree Lane*. Independent, 2019.

[20] Jackson, B. *At Home on Apple Tree Lane*. Indep., 2019.

[21] Jackson, B. *Cortland Place on Apple Tree Lane.* Independent, 2021.

About the Author

Betty Whitaker Jackson writes Christian fiction, nonfiction, poetry, children's books, memoirs, and devotional guides. In the last ten years, she has written 25 books and is published in twelve juried anthologies. She won First Prize in the *LifeRich Reader's Digest Memoir* Writing Contest for her book "Rocking Chair Porch: Summers with Grandma." She blogs regularly at https://www.bettyjackson.net. The trailer for her book *Whispering Woods* can be viewed online at https://www.youtube.com/watch?v=Pz_MPtVkiKY. She thanks the Space Coast Writers' Guild for its encouragement.

AFTERMATH

diplomedia / Shutterstock.com

Many parts of the world were changed in the aftermath of the COVID-19 pandemic. Sadly, some things never change. On August 4, 2020, the Beirut Port and surrounding buildings in Lebanon were leveled due to a blast triggered by a warehouse fire. The building was unsafely storing ammonium nitrate for over six years. Oversight in the area was always lax; during the pandemic, it was almost non-existent.

There were "at least 218 deaths, 7,000 injuries, and US$15 billion in property damage, as well as leaving an estimated 300,000 people homeless." (Wikipedia)

AFTERMATH

"The COVID-19 pandemic changed the relationship between the market economy, state, and society in the G-7 countries and beyond. While economies collapsed due to the shutdown of broad swathes of the economy, the state and civil society have gained new significance in protecting people from the pandemic's effects. This dramatic shift has recalibrated the public's perception of the role of markets, government, and society in response to the worldwide shock."

Katharina Lima de Miranda and Dennis J. Snower

Global Working Paper #154

The Brookings Institution

April 2021.

THEURGY

Nicholas Kaplan

God awoke from a nap to a commotion in Heaven. It seems it was initiated by one particular seraphim angel and several cherubs. They were arguing about the future of one of their many creations and if the Devil (who worked on the other side of the universe, commonly known as the Dark Side) was up to something sinister.

The seraphim angel stomped around, waiving and fluttering the large, heavy wings with such force as to cause a wave within the Cosmic Wind. A cherub, one with a ukulele, plunked so hard on the strings that they broke. The resulting sound was dissonant and strange and unlike the melodic stream everyone in Heaven was used to hearing. God was upset.

It seems the seraphim was sure the Devil had concocted a plan to try out on the bright side of the universe and prove to everyone that, with suitable specimens, the laws of nature, reason, humanity, and everything else could be compromised. Then, with a whisper, an archangel asked, "Why do you think the people of Earth are convinced that they see visitors from another planet? These strange events are nothing more than a scheme at disillusionment and to confuse the truth. It is a prelude to an alternative belief."

The cherubs, always very kind and rather whimsey in nature, do not believe the Devil would violate the boundaries established many epochs ago. Why now? Their faith and pride prohibit them from questioning anything other than The Now. By the mere fact of this argument, they violate their *raison d'être.* Cherubs are to promote the essence of happiness, joy, and love. At least that's the way painters portray them. Their reasoning and beliefs are being tested and questioned. Could the suspicions of the seraphim prove correct?

***

God, picking up a copy of "The Record of TIME," reads about all the turmoil, destruction, and confusion of the unique creation. God recalls all the missteps made, like the Phanerozoic eon. There are constant annoyances and events such as floods, famine, plague, pestilence, war, and so on. Notwithstanding the great joy in those brief moments of serenity and peace, God must now evaluate all events that endanger the creation's tranquility and weigh the arguments of all angels, but their prophecies are bleak.

Two million-plus beings depart in one year without cause. The God creation named Amazon.com is split into several entities, with some divisions controlled by a government. All social media platforms become diseased with viruses and corrupted by exploitation from within. Artificial intelligence directs all manner of decision making, from defense to allocation of all resources upon the land, air, and sea. The food supply is diminished. After fifty years, there is still no viable solution to the electric battery disposal problem. This creates pollution for both animal and human populations. Governments are in a perpetual state of turmoil. Songbirds disappear. There is so much to come that the archangels, weeping and crying, stop their speculation to avoid angering God. They hide in Heaven.

Everything is different, yet nothing has changed. The Sun rises and sets even if we cannot observe the event. The Moon still dances across the void, pulling and pushing the seas in the rhythm of the cosmos, but anxiousness ripples through the thought processes of humanity. In this pandemic, the spectrum of human emotion runs from scenarios of fear to fatalistic paralysis. Despite the advancement and accomplishments of science through the ages, humans have often been crippled and humbled by inconspicuous existences. These are minuscule forms, evolutionary derived, that can eradicate all life.

The survivors of this pandemic experience total disillusionment and a deep skepticism of politics, religion, and their future. With a stiff upper lip, the eight billion people of the creation march on with fear and trepidation. This assessment will spill over into the spiritual, emotional, and psychology of everyday life. Although humans want to put this behind us and look towards the future, a lingering and annoying remembrance of outright lies, delusions, and mismanagement will poison all attempts at solidarity. The sanity of the world's population is being tested.

***

In a humorous way, God seeks out advice from one of many guardian angels assigned to souls at birth. They are to protect and guide the inhabitants in life decisions and protect them from harm. The angel chosen had been a problem for many years. Always complaining about the "workload" and being late for angelic meetings, not bringing the heavenly donuts, etc. God, however, decided to inquire as to how things were going before making any decisions. This angel was consulted because it was one of the first selected to become an angel in Heaven and had completed many complex cases.

From the angels' observations, the usual events were transpiring with a slight deceleration in pollution damage, less economic disparity, and a renewal of faith. However, this guardian was unlike the group (commonly known as a flutter of angels) in that it encountered thoughts of hope, love, and tranquility predominating the consciousness of the inhabitants. With this drastic change in direction, humans begin to embrace contentment, kindness to others, and protecting life and the environment. The regular bombardment of politically-directed news diminishes. Leaders of nations unite in a consolidated effort to "save the planet." They reject false profits and their futures. Families are reunited after many years of isolation due to the pandemic virus. Schools are open, socialization has become the norm, and advanced technology is only developed and used to better the planet. Events become more manageable in a very short time, and songbirds start to reappear.

God, not wanting to destroy everything and start over, begins deliberating. Humans are not perfect; I made it so. People, since the beginning, have been frightened and amazed at the events I have presented to them. I do this to keep them on their toes. It is necessary to adjust their attitudes and mindfulness of where real power exists. Some are overcome with awe, yet others try and diminish the accomplishment and take credit. They spend their precious time seeking the origin of this power and to what end.

Frustrated, God decides to allow the known universe, with all its misgivings and problems, to continue. Sometime in the future, a review ("Rapture") will be scheduled and a final decision made, but for now, humans are free to affect their state of being. The guidance shall continue to be provided by the appropriate angelic force, and accommodations shall be made for the additional souls headed towards Heaven.

Having decided, God again reclines on a cloud, listening to the angelic cherubim play heavenly tunes on their ukuleles. Then, picking up a new "Record of TIME," he reads an article on "Daylight Savings Time." Meanwhile, on the Dark Side of the universe, the Devil and his minions, upset with the performance of a particular human (45th President of the USA), starts in motion a very, very, very large rock, which they send towards Earth.

In a cold sweat, Nicholas wakes up.

About the Author

Nicholas Kaplan is a first-time author living in Florida. Since the 1960's, he worked and traveled throughout Europe and the Middle East. Now retired, he has written articles and essays for the Brevard Scribblers anthology *Driftwood,* and is a member of the Space Coast Writers' Guild. His self-published novel (under the *nom de plume* of Nicholas Taylor*)* entitled *The Long Game* is a 21st century story about espionage and government corruption, and is available on Amazon.

AFTERMATH

appledesign / Shutterstock.com

The "end of days" are often seen as apocalyptic. Throughout history, the aftermath of pandemics is portrayed as periods of death and destruction followed by change and rebirth. Thankfully, we only had to wear face masks during COVID-19; gas masks were quite rare.

AFTERMATH

"Fear none of those things which thou shalt suffer: behold, the devil shall cast some of you into prison, that ye may be tried; and ye shall have tribulation ten days: be thou faithful unto death, and I will give thee a crown of life."

Revelation 2:10 (KJV)

Somatosensation

Elayne Kershaw

Note: An earlier version of this story first appeared in *Written in the Sun: Brevard Scribblers Anthology 2021*. This updated version is included here with permission.

"Somatosensation" is a companion piece to the story "Friendship and Love in a Post-COVID-19 World," which appears in the next chapter. "Somatosensation" is set entirely in England and offers a metaphor for a world devoid of human touch.

Taken together, the two pieces of writing anticipate, in different formats, the nature of human communication in the aftermath of COVID-19.

"One of the primary ways our bodies trigger oxytocin production is specifically through physical touch. Without oxytocin, we start losing out on those very feelings and bonds that make us uniquely human."

Nadia Ameri, 2020

Introduction

There were three legally enforced lockdowns in the UK between March 2020 and January 2021. These restrictions on people's movements, intended as a deterrent to the spread of COVID-19, affected relationships at every level of society negatively and positively. Many people were happy to work from home and spend more time with their loved ones (especially their children), but others, particularly those who lived alone or in remote, rural communities, reported an increase in their sense of isolation. While some welcomed the convenience of Zoom interactions, others sounded the alarm bells about the devastating impact of the loss of physical contact with other human beings.

January 31, 2020

Alex was reading the newspaper. The first two cases of coronavirus had been confirmed in the UK. As his eyes scanned the story's main points, he looked up every so often to check on Megan. He could just about see her in the pale, early morning sunlight through the streaks of condensation trickling down the patio doors. She was feeding the birds. She had wrapped a fuchsia-pink blanket around her silk dressing gown and, clutching at the collar with one hand to keep it in place, she was braced against the wind, scattering nuts and seeds. There was no boundary wall at the end of the garden, just a wide-open view of the plowed fields beyond. The deep umber ridges were frost-bitten with no trees or hedgerows to soften the sting of winter. Against this unforgiving backdrop that stretched unbroken to the horizon, she was a tiny spot of color. Brave but frail, like the half-starved robins that fluttered around her.

He turned the paper over and set it down on the arm of his leather chair, sliding his fountain pen over the now-crinkled pages, ready for the crossword, which he would do in the afternoon with a cup of tea and *Dickinson's Real Deals* for company. A burst of cold air swept through the house as the kitchen door opened abruptly and then slammed shut. He could identify every sound she made: the scrape of her feet on the braided-rope doormat; the swish of fabric sliding down her body as she disrobed; the crackle of her hair as she swept it back from her face; the sigh of discontent, drawn deep from her lungs, as she exhaled.

Eventually, she appeared in the doorway, shivering slightly but drawing warmth from their enormous ragdoll cat, Velvet, who was now draped across her arms, all four legs dangling.

"Would you like me to turn the central heating up," he asked her, indicating the thermostat with an encouraging sweep of his arm, "or just stoke the fire in here?" She looked in the direction of the fireplace and shook her head.

"No," she replied, "I'll be fine. I have to go into town shortly." She turned on her heel and made her way to the stairs. He followed her and reached out to stroke Velvet, who was staring at him from over her shoulder. As he made first contact with the whiskery tips of Velvet's outermost layer of fur, a charge of static stabbed his fingertips. He drew back, pushing his hands deep into his trouser pockets. She ascended the stairs slowly. He resisted the urge to help her, which he still felt, even after all this time. Instead, he stood guard at the bottom, ready to catch her if she fell.

"Have you been following the news?" he called up to her when she finally reached the top, "about that virus discovered in China? Unfortunately, it looks like it's reached our shores."

She paused.

"I did see something about it," she said vaguely, before continuing the journey to her bedroom.

March 14, 2020

The telephone rang. It was Jamie, their son, calling from Los Angeles.

"Hi, Dad." Alex loved him not just because he was their only child, or that he had been a model son who was now a successful vascular surgeon on the other side of the Atlantic, but because he had never pressed his case for the use of Internet calling, which would grant them the gift of vision as well as sound. Alex often thought it was a miracle that Jamie had also been endowed with emotional intelligence. Today, he wondered why he sounded so alarmed.

"What's up, son?" he asked him.

"Dad," he said, "Mike Pence has just announced that the US is extending its European travel ban to include the UK. It's coming into effect in two days' time."

"Well," Alex replied carefully, "I suppose that's sensible."

"I agree, but do you realize what that means? You can't come here, and I can't go there, for who knows how long."

"Let me tell your mother first," he said, "and then she will call you back."

He found her in the greenhouse, cleaning her gardening tools and preparing for spring planting.

"I need to tell you something, my dear," he said as he

shuffled his feet nervously. "It's about Jamie and Easter."

She didn't call Jamie back. He heard her crying quietly in the bedroom, and his heart ached for her.

March 15, 2020

Alex drove Megan to her doctor's appointment. He was grateful it had not been canceled. So many other GP practices had abandoned all face-to-face communication with patients, fearful of transmission, despite the government's controversial aim of achieving herd immunity. When they drew up to the traffic lights, he studied her profile. When they were very young, he told her that she would always be beautiful, and every time he looked at her, he was struck by the intense blueness of her eyes; the long, graceful shape of her nose; the striking angles of her face. She was wearing a musky perfume with orange notes. She sat perfectly composed in the passenger seat, with the seatbelt pulled tightly across her waist. This was the only activity they had shared for longer than he could remember.

During the monthly visits to the surgery, he would sit in the waiting room and thumb through crumpled pages of the *Reader's Digest* when she was called in to see the doctor. She never disclosed what was discussed or decided, but sometimes she would be flushed and trembling on the way home. He switched the radio on, just in time to hear Matt Hancock, the Health Secretary, announce that every UK resident over the age of seventy would be told within the coming weeks to self-isolate for a very long time to shield them from the virus. Everyone else was going into lockdown. She leaned forward as far as she was able so that she could reach the dial and turned the radio off again.

"We'd better do a shop then," she said, "we'll need to stock up."

March 16, 2020

They went to Waitrose the following day. Megan went to the pharmacy section to collect her prescription while Alex wheeled the trolley patiently up and down the aisles, following the newly painted green arrows. At one point, he went the wrong way and collided with several angry shoppers, who hadn't made that mistake, and who kept jabbing their fingers at him and then back at the arrows on the floor. They made him reverse back, out of their way.

All the hand gels, sanitizers, toilet rolls, and pasta were gone. Row after row of shelves had been stripped bare. When he met Megan at the check-out, with the few items he had been able to grab, she was clutching a huge paper bag.

"The doctor anticipated the situation," she said drily, "and changed my prescription. I've got a six-month supply." She dropped the bag into the half-empty trolley. He felt ashamed, as if he had let her down somehow.

"We'll have to keep coming back," she announced. "We can't starve, no matter what Matt Hancock says."

"We've gone into lockdown, dear," he reminded her, "and you and I have to stay indoors. We'll have to order food to be delivered."

"I can go in the garden," she declared, "I can't stay in the house ...." she paused and then added, "all the time. I'd go mad."

He felt and heard the pause and knew that what she had deliberately omitted were the words "with you." He noted that at least she wanted to spare his feelings, even if she couldn't spare her own.

April 30, 2020

Megan and Alex were eating their evening meal in the dining room. Megan had conceded that to keep the number of supermarket deliveries to a minimum, it made sense to order food for shared – rather than separate – meals. Some compromise had been necessary to achieve this victory, which was how Alex viewed it. He had traded his carnivorous diet for vegetarianism, which was her step down from an herbivore diet. With very little else to do all day, they had peeled and chopped onions, cabbages, leeks, potatoes, and anything else that would make hearty soups and non-meat stews. They had started loading the dishwasher together and even managed to joke about it. As winter had given way to spring, they had switched to salads and fruit-based desserts. Alex had lost weight.

In a reckless moment, Alex suggested a glass of claret. He was astonished when she agreed not only to that but to drink it with him in the lounge, in front of the television. He pulled the cork out of the bottle with a flourish. They sat in opposite chairs, but they were still in the same room, and she was smiling.

"Shall we watch the evening news?" he asked her.

Boris Johnson announced that the UK was past the peak of the COVID-19 outbreak and, although the country could not risk a second spike, he was recommending a comprehensive plan for easing the country out of lockdown. He didn't mention the over-seventies, like Alex and Megan, who had been living like prisoners for a month.

“That’s wonderful news,” she declared. “I can get you new clothes and do a big food shop.” She held out her glass for a refill. Alex saw that her cheeks were red.

“Should you?” he asked tentatively.

“Oh, one more won’t hurt,” she laughed.

He poured the wine into her glass, watching the reflected lights of the chandeliers dance on the edges of the finely cut crystal. Their eyes met over the sparkling rim. He poured himself a glass and held it out to hers.

“Let’s have a toast,” he offered.

“What shall we drink to?” she asked him.

He thought for a moment.

“To togetherness,” he replied, “and new beginnings.” She nodded, and emboldened by this, he coiled his arm around hers so that they could drink from each other’s glasses. Then, very gently, he took her glass and his and placed them down on the table.

***

In the days, weeks, and months to come, he would

replay the next moment over and over in his mind, wishing that he had read the situation differently. Wishing that, like his son Jamie, he had emotional intelligence. When he relived the scene, he saw it as if he were watching it from the outside in, as if he were watching television. He saw Alex slip his arm around Megan's waist and draw her in close so that he could kiss her. It was over in seconds. She pushed him away with such force that he fell against the table and knocked the wine glasses over. The wine poured onto the carpet like blood, and the lounge looked like the setting of a murder. She stumbled from the room and left him, motionless, to survey the damage done.

June 19, 2020

Something woke Alex. He sat up in bed. It was still dark. He listened intently. He usually slept his full nine hours, like clockwork. It was hot, and the window was open. He leaned out over the windowsill, thinking he would catch the loud wails of mating cats or the distant hum of early morning traffic. But all was quiet.

He got up and padded over to his bedroom door, where he placed his ear against the warm wood, swollen by the summer temperatures. He heard something. Opening his bedroom door gingerly, he crept over to Megan's room, anxious not to wake her unnecessarily. She was coughing.

He went downstairs to make her a hot lemon and honey drink to soothe her throat, but as he stood waiting for the kettle to boil, he heard the cough mutate into a terrible wheezing sound. He ran up the stairs and threw her door open. She was lying in amongst a mass of crumpled sheets. Her mouth was open, and the most terrible sounds were coming from the back of her throat. She was shivering and clutching at both her head and her chest.

He didn't dare touch her. Instead, he picked up the phone at the side of her bed and dialed NHS 111. When a voice answered him, he spilled out his words, too many, too quickly. She urged him to slow down, calm down, and try again. When he described what was happening to his wife, she told him that he had to dial another number and that when he did, the call would be automated, and he needed to answer all the questions as clearly as possible. His responses would determine what would happen next. But, of course, he couldn't answer all the questions because they had barely spoken for weeks. All he knew for sure was that she had been out of the house on one or two occasions, but only to go to the supermarket and go to Marks and Spencer's to buy him some smaller trousers and shirts. He couldn't say whether she had been feeling unwell in the previous days.

The ambulance came for Megan in less than an hour. The paramedics were swathed from head to toe in protective clothing. Unfortunately, he wasn't allowed to accompany her.

June 22, 2020

Alex turned the lounge into a makeshift bedroom where he sat and waited for contact from the hospital. They had instructed him not to clog up their over-stretched phone lines. Instead, they would call him with updates. Occasionally he would doze off and then wake up again with a start, his head pounding, and check for voice messages. He tried not to torture himself with thoughts about the irony of her condition: that she should succumb to this virus at a point in time when the distancing rule of two meters was going to end, and that the UK had recorded its lowest number of new cases since the beginning of lockdown. The government was so optimistic that an end was in sight that they were scaling down their daily briefings.

June 26, 2020

Megan was placed on a ventilator. Her condition was so serious that he had to brace himself for the possibility that she might not survive. Despite this, he still wasn't allowed to see her. He called Jamie, who did not mention that some patients could talk to their loved ones on iPads or Tablets and blow kisses to them and promise them they would beat this thing and be home soon. Unfortunately, Megan was too ill to talk.

Alex prayed. He didn't believe in God, but he prayed anyway. He was willing to try anything, and he could do very little else. He gathered up all the photographs he had of her and placed them in his line of sight. Then, pulling Velvet onto his lap for companionship and comfort, he analyzed them forensically and saw, for the first time, the subtle nuances, the changes in her demeanor that they captured. Month by month, year by year, the physical distance between Megan and Alex had become greater. In the last ten years, she stood completely apart from him, her hands clasped firmly together and her eyes wide open but unfocused. He confronted the knowledge that he would probably never pinpoint the moment or the catalyst that triggered the change in her. He remembered, again, how his clumsy attempt to reconnect with her had backfired, provoking suppressed anger to bubble to the surface and harden, like cement, between them. He buried his face deep into Velvet's fur and cried long and loud into the night.

September 17, 2020

Megan spent four weeks on the ventilator, during which time she had to have a tracheostomy tube inserted below her vocal cords. The ventilator, brutal though it was, saved her life and stopped her from fading away. She was weaned off it gradually, and when she was finally able to breathe on her own again, she was moved out of the intensive care unit and into an isolation room. She suffered pneumonia three times and protracted periods of delirium, for which she was given extensive physical and occupational therapy. Nevertheless, the medical team was cautiously optimistic that she would, in time, make a partial if not complete recovery.

Alex received the phone call he had almost lost hope of receiving. He would be allowed a very short visit, subject to a negative COVID-19 test. He would have to wear a complete set of protective clothing and couldn't enter her room. Before he went to the hospital, he called Jamie, who was already on his way to the airport.

***

He shuffled up to the small window, his movements limited by the weight of the additional layers of clothing. The window measured less than two feet square. Despite his height, he had to stand on tiptoe to see through it. His calf muscles contracted sharply as he concentrated on the effort to stay in position. He leaned his chin on the window ledge and gripped it with his gloved hands. The room was surprisingly minimal. She was connected to a monitor and a drip, but other than that, it was just her, in the middle of a bed, with the starched, white sheets folded pristinely across her body, concealing the network of wires. She was on her back with her head facing the ceiling. Her eyes were closed. The monitor pulsed at regular intervals, spilling neon-blue data across its screen.

He hadn't been allowed to send flowers or cards, so the only communication she had experienced in three months had been with anonymous medical staff. He tried to imagine the tube being pushed into her throat, making her gag and panic, seeing strange, goggled eyes staring down at her, and sensing unfamiliar hands stabbing her with the needle that dragged her back down into oblivion. Buried alive in that halfway house between consciousness and unconsciousness, life and death, she couldn't have known how hard he had tried to transmit his love to her across time and space.

She made an almost imperceptible movement, and he saw that her hands were not clenched tightly as they usually were. Instead, they were outstretched, and he thought, though he wasn't sure, that she was trying to move one of her fingers. He pressed his face against the window and tentatively tapped on it. Nothing. He tapped harder, then harder still. Finally, she turned her head slowly in his direction and opened her eyes. He saw a flicker of recognition. She tried raising her head off the pillow, but she was still too weak. He looked around him, checking to see if there was anyone in the corridor, but he was alone. Quickly, he ripped his glove off and placed his bare finger on the window. He pressed so hard that the end of his finger went white and numb. The corners of her lips moved upwards into the faintest of smiles, and she lifted her finger a little higher. Despite the wall between them, they were touching at last.

About the Author

Elayne Shershaw was born in London and later lived in Cornwall, England for twenty years before moving to Florida in July 2018. She began her career in journalism before lecturing in English, Creative Writing, and Film and Media. Now retired, she is working on a collection of short stories about human communication in the digital age. She scuba dives in the warm waters off the southeastern Florida coast to produce underwater art films for her YouTube Channel *Mediterranean Blue*

Films. Elayne is currently in charge of marketing and publications for the Brevard Scribblers. She lives with her husband Alan and their two cats, Coco and Chloe.

AFTERMATH

Maksim Shmeljov / Shutterstock.com

Modern love in the aftermath of COVID-19. We prefer to meet outdoors. We still practice social distancing (at least on the first date). We still wear masks. Limited physical human contact goes against our very nature, but we're forced to accept the situation for now. Let's hope the bench gets shorter in the future.

AFTERMATH

I've seen people turn their heads
And quickly look away
Like a newborn baby
It just happens everyday

I look inside myself
And see my heart is black
I see my red door
I must have it painted black

Maybe then, I'll fade away
And not have to face the facts
It's not easy facing up
When your whole world is black

:
:

If I look hard enough
Into the setting sun
My love will laugh with me
Before the morning comes

Excerpt from *Paint It Black* by The Rolling Stones

From their 1966 album ***Aftermath***

FRIENDSHIP AND LOVE IN A POST-COVID-19 WORLD

Elayne Kershaw

Note: "Friendship and Love in a Post-COVID-19 World" is a companion piece to the story "Somatosensation" that appeared in the previous chapter. Taken together, the two pieces of writing anticipate, in different formats, the nature of human communication in the aftermath of COVID-19.

"Friendship and Love in a Post-COVID-19 World" focuses on the dynamics of human relationships and makes comparisons between England and America, which the author has been traversing for the last several years.

I found my husband-to-be (Alan) on Facebook in July 2016. Even though he was now living in Florida, I recognized him as my first serious English boyfriend in the seventies. In that century and decade, relationships of all kinds were often instigated, perpetuated, and ended by telephone and letter. Our relationship, however, was sabotaged by our respective parents, who had an explosive and irresolvable argument. They banned us from seeing each other. His family moved to a new town, as did mine, and we were lost to each other. Fast forward forty-plus years, during which time the world experienced a digital revolution and became a global village, in the famous words of Marshall McLuhan. We began a new, virtual relationship that we conducted online through the magic of technology.

I had searched for him before, when I first joined Facebook in 2008, but I don't think their "crawlers" were as sophisticated in the early days, and I quickly became side-tracked by the hundreds of "friends" who added me to their network. These were the kids I grew up with: teachers, girlfriends I had lost track of, neighbors, work colleagues, bosses, employees, students, art and writing club friends, and even hospital bedfellows. These people supplemented my immediate circle of family and friends but did not replace them. I treated these two camps very differently. Friends on Facebook received "Likes" and "LOLs" and the occasional joke or funny picture, which were reciprocated. They wished me a happy birthday once a year and occasionally suggested that those of us in a sub-group should all meet somewhere central. We never did. Facebook lets you keep track of what people are doing, in almost real-time, with the scroll of a mouse, so that you feel you are a good friend to all of them – from a distance.

Meanwhile, I saw my "real" or "proper" friends and family, as I called them, in person as often as my heavy work commitments would allow. This involved hours and miles of travel on weekends. Still, we were all committed to making an effort because we knew, without even saying it, that it was the physical togetherness that sustained our relationships.

By 2016, I had added Skype, WhatsApp, Twitter, Instagram, LinkedIn, and Messenger to my collection of social media platforms. I had many of the same contacts on most of these platforms, but the target audience of each of these was specific, so this allowed me to differentiate between, say, artistic friends on Instagram and business connections on LinkedIn. In addition, the method of communication varied between texting, voicing, and video-calling, which in turn determined the content and depth of our interactions.

Alan and I used Messenger to get to know each other all over again because it allowed us to stay connected for twenty-four hours if we so wished, depending on the quality of our Internet access at any given time and place. We calculated that we must have been online for hundreds of thousands of hours between him first calling me on Messenger (in response to a tentative "Is this you?" message I sent him) and me moving to the US.

We wore headphones and talked to each other when we were driving to work, going to the shops, walking the dog, doing the garden, and even when one of us slept, we stayed connected and listened to the reassuring sound of each other's rhythmic breathing. We sent long text messages and shared our favorite love poems when we couldn't speak or listen. The constant theme of our conversations was how fortunate we were. Because of technology, we could be together in this way, despite the miles and time zones that separated us. But we still yearned to be together "in the flesh."

***

We were married in England in 2017; my American visa was granted in June 2018. While we waited for this happy event, he flew to England every three months or so. He hadn't lived there for over twenty years, and while he was struck by the changes that had taken place in his absence, he was also delighted to find that many customs and traditions remained intact. One of these was the habit of dropping in on friends or family unannounced and putting the kettle on. Such visits were nearly always welcomed, and an hour or so could easily slip by unnoticed while the world was put to rights. As far as the British are concerned, there is no problem that cannot be solved over a cup of tea.

Alan made his last trip over to me at Christmas 2017. As I drove him to the airport for his flight back, he asked me if I was prepared for the culture shock that awaited me when I would finally join him. I didn't understand how there could be significant differences between two countries that shared the same language and history. He smiled and told me I would see what he meant in due course.

***

When I flew to Florida in 2018, I left behind my family, friends, social networks, and my job as a teacher in a sixth-form college. I planned to become an established writer and artist (two pursuits that had been on the back burner for most of my working life). I hoped to meet a whole new network of creative people with whom I could collaborate.

After the initial "settling in" period, in which I sorted out a lot of paperwork, Alan started taking me to his business network meetings to meet the people he worked with locally. This was where I first experienced some of the significant differences between America and Britain in the ways people interacted socially. When Alan introduced me to an individual or a group of people, they invariably asked me, as an opener, what I did. It seemed that people here were defined by their job or career, and at first, I was embarrassed and defensive. I said I would be looking for employment in due course and hoped to teach again. Eventually, I stopped saying this because I had no intention of ever teaching again. I developed the confidence to speak the truth, which was that I was looking for an art studio and freelance writing opportunities.

The people with whom Alan networked held regular breakfast meetings, but these were very formal, and I decided that these were not appropriate venues for me to try to make new friends or reach out to people, which I learned was the correct expression to describe what I was attempting. They also met in local bars and restaurants, where everyone relaxed.

We were invited to after-hours events, meals with other couples, the odd barbeque on the beach, and several promotional events when someone was launching a new business premise or a new product. We were rarely invited to anyone's home. Instead, social arrangements were scheduled, just like business appointments. There was often an underlying agenda in which the get-together was conducive to a mutually beneficial enterprise.

I tried inviting some of the women I met at these events to my home but was always politely declined and, I had to admit, for a good reason. They worked long, hard hours, and on rare days off, they took care of shopping and medical appointments. As a result, I felt lonely and homesick. I alleviated this by staying in contact with friends and family in Britain online and by flying there as often as I could. On these visits, I poured my heart out about the fact that I was never going to be able to turn up at someone's house in Florida without an appointment and ask to come in for a drink and a chat. It just wasn't done. If you wanted to meet up with someone, you had to schedule a time and a place beforehand and be prepared for it to be re-scheduled, often at short notice. The meeting had to have a reason of some kind, and it had to take place in a diner or a restaurant of some sort, during someone's lunch break. The concept of just getting together with a friend for a meandering chat with no agenda and no time limits didn't seem to be feasible.

***

I knew that the onus was on me to adapt to a different way of life, and I tried very hard to do so. What helped me adjust was a series of serendipitous events. First, through my love of scuba diving, I made friends with a wonderful man (who was American and became my Florida dive buddy) and his lovely wife. They welcomed me with open arms into their home, bucking the trend. I encountered a warm and hospitable diving community in Palm Beach through him.

A short while later, I met a large group of British ladies who regularly convened in each other's homes to support each other and reminisce about life back in Blighty. I was drinking tea and chewing the cud again.

My dive buddy also introduced me to a group of writers who met twice a month. Here were my creative "brothers and sisters." It was beginning to come together. I was enjoying the best of two worlds. The only irony was that I was now conducting my relationships with my original friends and family in Britain electronically while "hanging out" with my new American and Brits-in-America friends face to face. Then COVID-19 happened.

***

Britain went into lockdown very quickly. People retreated indoors, creating ghost towns everywhere. All but essential businesses closed. Hospitals canceled operations. Flights were grounded. People exercised alone for limited periods of time within a short radius of their homes. The only vehicles on the roads were permitted delivery trucks and the emergency services. Business and social communications moved swiftly onto platforms such as Zoom, as people rallied around and found ways of overcoming these stifling restrictions.

In America, the response varied from state to state and city to city. Some governors ordered full and extensive lockdowns, while others oversaw the closure of many public-facing businesses. A lot of people elected to place themselves in isolation. Mass gatherings and sports events were suspended. Beaches were shut. However, there was a sharp divide in attitudes to the coronavirus and how to deal with it, which some people attributed to POTUS. Those who followed his views (e.g., that the economy should stay open, and that the severity of coronavirus was being exaggerated) refused to wear masks or socially distance. Those who did not follow his views covered their nose and mouth carefully if they had to venture outside.

Britain made a fatal mistake and relaxed restrictions in the summer, believing the worst to be over. People started to gather again, in large numbers, mainly to attend rallies and marches. The event that provoked this foolhardy behavior took place in America. A young black man, George Floyd, was killed after several police officers, in the act of trying to arrest him, knelt on his neck for over eight minutes and killed him in the process. The Black Lives Matter movement gained momentum in response to this, not just in America but around the world, and nowhere more so than in Britain, which was still trying to come to terms with its racist history. The statue of Edward Colston, a notorious slave trader, was pulled down in Bristol, the port that had handled the transfer of kidnapped enslaved people from

Africa to the New World. Britain was already painfully divided between Brexiteers and Remainers as the date for leaving the European Union drew nearer; now, there was another reason to widen the national wound.

I saw the effects of this on social media. Friends I had known for years who I had always thought of as being rational, intelligent human beings started to spread conspiracy theories and misinformation about, for example, the BLM movement, the coronavirus, Antifa, and the "Radical Left." If I challenged them, however nicely I phrased my responses, they unfriended or blocked me. It was so quick and easy to do that. I thought back to the seventies when you had to use paper and pen, spend time on your composition of words, address an envelope, buy a stamp, and make a trip to the post office before you could cut someone out of your life forever. The steps necessary to send physical mail gave you enough time to calm down and ask yourself if the issue you had disagreed over was worth sacrificing a friendship years in the making.

By November, Britain was back in a total lockdown. The approaching winter weather had already impacted people's enthusiasm for marching and fighting in the streets, but the damage was done. Families and friends were irreconcilably divorced from each other, having decided that they had never really known each other at all. The phenomenon of ruthlessly dueling and dispensing with loved ones over ideological differences started to spill over into other aspects of online communications. A video of a routine council meeting, conducted via Zoom by a small group of elderly parishioners from a sleepy village in Cheshire (Handforth), England, erupted into a loud and ugly shouting match that was replayed on mobile devices around the world and achieved millions of hits on YouTube. The voyeuristic interest in this sad altercation (in which a few men ganged up on one woman) was indicative of an insidiously creeping and growing social problem. The more physically isolated people became, the more willing they were to become either reclusive or combative online. In the latter case, "keyboard warriors" replaced anonymous "trolls" – proud and ready to own their divisive comments. The latest bone of contention is whether or not to vaccinate. People have become polarized into "pro or anti-vaccination" stances. Both armies believe they are right and have their weapons to prove it. There are no peace talks and no compromises. The battles are unforgiving.

***

When historians write this period of history, they will undoubtedly describe it as the time when the nature of humanity itself was at stake, and we were in danger of regressing to the dark ages. In the novel *The*

War of the Worlds by H.G. Wells, civilization is almost wiped out by Martian tripods until the extra-terrestrials themselves are brought down by a simple pathogen (or germ) to which they had no immunity. The narrator observes both the best and worst of human behavior during the terrifying occupation and makes pointed comparisons between the acts of social cooperation that saved lives and the acts of self-interest that cost lives.

If we live in an alternate version of Wells' narrative, we've reached the point of no return. Many of us are anesthetized and desensitized to human contact. We have learned to live separated and isolated – by the coronavirus and our political views. We have become insular and either withdrawn or socially inept.

When we emerge from our cocoons in the post-COVID-19 era, we will either recognize the loss of something precious and search to recover what we abandoned or cut our losses and redefine the meaning of humanity.

About the Author

Elayne Shershaw was born in London and later lived in Cornwall, England for twenty years before moving to Florida in July 2018. She began her career in journalism before lecturing in English, Creative Writing, and Film and Media. Now retired, she is working on a collection of short stories about human communication in the digital age. She scuba dives in the warm waters off the southeastern Florida coast to produce underwater art films for her YouTube Channel *Mediterranean Blue Films*. Elayne is currently in charge of marketing and publications for the Brevard Scribblers. She lives with her husband Alan and their two cats, Coco and Chloe.

AFTERMATH

Photographee.eu / Shutterstock.com

One of the more unfortunate outcomes in the aftermath of the COVID-19 pandemic was a surge in breakups and divorces. Prolonged lockdowns, the loss of jobs and the subsequent loss of income and identity, and the lack of external social interaction (except for Zoom) took a real toll on personal relationships.

AFTERMATH

While many families have enjoyed spending more time together during the pandemic, there are some relationships that have failed to thrive during a period of unprecedented upheavals and uncertainties.

From arguments over Covid rules and restrictions to disagreements over whether children should be vaccinated – and even disputes between families and friends over the very existence of the virus – have seen relationships pushed to breaking point during the pandemic, according to family law experts and psychologists.

"Marital conflicts have been on the rise since the pandemic. I have noticed a rise in the number of clients seeking a divorce. I get three to four inquiries a day for my services, whereas prior to Covid the inquiries were much less," New York City divorce lawyer Lisa Beth Olde told CNBC.

Arguing with your partner over Covid?
Holly Ellyatt
Health and Science, CNBC.com
January 21, 2022.

https://www.cnbc.com/2022/01/21/covid-has-put-pressures-and-strains-on-relationships.html

The Factors, the Problem, and the Product

Nanda Klein

My body was so hot I could feel the sweat rolling down the back of my thigh. However, the tip of my fingers got cold in seconds. I hoped not to get sick after walking out of the Miami heat right into what felt like a blizzard.

No, I didn't get off a plane in New York City. Instead, I had just stepped inside a retail department store with the air conditioning blasting so high it made my hair blow like I was Beyoncé on stage.

The wet sanitizing wipes weren't helping warm my hands at all. Once I finished disinfecting the shopping cart handle, I put on my sweatshirt and prepared for another shopping adventure amid a pandemic.

Holding the handle with a tight grip, I hurriedly pushed the cart toward the battleground. Dealing with people on the verge of a psychotic attack made me anxious. Any mistake could unleash the rage of an individual unprepared to deal with the lack of social distancing many failed to keep. For example, I recently witnessed a lady yelling at someone for grabbing produce with their bare hands.

What was the person supposed to do? Buy rotten tomatoes just for the sake of not touching anything else? Besides, gloves were even nastier after touching many surfaces since they could not be sanitized. Regardless, after that occasion, I began to handle everything with a clean plastic bag taken from the produce area. I didn't have the time or energy for any more drama. The pandemic was already taking over all the space and time I had.

My peripheral vision caught sight of movement. I could not hear the distress due to the loud music playing in my ears. My AirPods had become my faithful allies; music helped me isolate myself from the sense of chaos whenever I headed out of my house. My anxiety levels skyrocketed just with the thought of going grocery shopping.

Once I looked, I noticed that Luis, the store manager, was trying to get my attention by waving his arms in the air. It was the first time I had seen him since an incident with one of his workers months before the pandemic. I smiled and waved back at him. I paused the music. "After all, I am not that much of a crazy lady, huh?!" He shook his head with a grin as I left his sight. "You have a good day!" His voice followed me around the corner.

Despite not knowing him personally, as a customer, I can attest that Luis is an outstanding manager. He proved it on many previous occasions, including the last time I had seen him when I stopped by to purchase a few grocery items at his store, a place known for being a tourist hot spot.

When I first moved from Brazil to Kentucky, I was constantly ill, partly because I didn't care much about my health. Also, as my doctor explained, it is common to get sick when moving to a different place since the body's immune system is not used to certain viruses to which the locals are resistant.

In addition, it made sense that travelers could be carrying it, spreading viruses and bacteria during their trip. With that in mind, particularly in Miami, I was careful. As a mother, it didn't hurt to do my part to avoid contracting an illness.

Therefore, during a trip to the store, after sanitizing the shopping cart handles and shortly after grabbing a few food items, I began to look at shoes for my daughter. While I scanned up and down the aisle for a specific kind, I noticed a worker nearby, dusting and removing products from the shelves. As she neared, I glanced over at her and headed her way. "Sorry! Let me get the cart out of the way."

Nevertheless, before I reached the cart, she responded in a dry, irritable tone, barely moving her lips, indicating that she would move it. Then she pushed the cart toward me. When my eyes met the yellow duster lounging on the surface of the handles between her hands, the greyish tone of the microfibers seemed to chant, *"Virus! Virus! Germs! Germs! Virus!"*

Her demeanor had already upset me. When I treat someone with kindness, the least I expect in return is good manners. The discourtesy in the tone of her voice had already made me uneasy. Now, her impatience escalated my frustration to a new level. I sighed.

Her stare, along with her disapproving duck lips, clearly announced her attitude. I could tell I had just ruined any chances of qualifying for customer of the month. *Oh, wait. There is no such thing!*

I'm all for courtesy, but if there's one thing I refuse to do, it's to walk away from a bitch contest. If one wants to be a bitch to me, come right over and let me lecture you how to be a real one. I can be the nicest person you ever met, but step on my toes, and I can be the craziest person who ever crossed your path.

"Great! Now I have to wipe my cart again!"

"Oh, my God! You are *sick*!" she declared with her mouth half open.

"Can you please get me some wipes to disinfect the cart again?" I pointed at the cart, moving my index finger from side to side with a hint of assertiveness in my tone. Now, the duck lips belonged to me.

Noticing her stubborn stare and inert body, I made a 180-degree turn and walked straight to the customer service counter, with her following, asking for Luis.

Luis showed deep gratitude for my comments, good or bad. In addition, he was exceedingly apologetic and very appreciative of my feedback. "I understand she is entitled to make whatever judgment of me she might desire, but it is preposterous to say such a thing to a customer, no matter how much of a lunatic I might seem to be," I said. I thanked him for his time and headed out of the door empty-handed.

***

Although I am uncertain of what ensued afterward,

that was the last time I ever saw her, which doesn't concern me a bit. My point is, like when we begin to learn the concept of multiplication in childhood, factors are given to us. Our task is to solve the problem by finding the product.

Isn't life equivalent? The factors are the circumstances we face. Resolving the problem corresponds to the choices we make. Finally, the product is nothing but the consequence of our choices. If she did get fired, she had brought that upon herself, and I certainly do not waste my time worrying about it.

In view of the pandemic, by no means is it my intention to imply that those who contracted the virus should be blamed. Instead, I wonder how we can prepare for a future global virus if we, as individuals and active members of society, cannot deal with unprecedented circumstances without panicking or overreacting. What prevents us from progressing on a personal level?

Here is a perfect example: when I walked into the store, I intended to purchase a few items I needed. Unfortunately, once I encountered a trivial challenge, I forgot what I wanted to accomplish because I got stuck on a foolish problem.

Nevertheless, I walked away empty-handed after solving the issue even though I knew better. While I believe I have the concept of emotional intelligence deciphered at the age of thirty-eight, proving a point seemed more relevant at the time. I got stuck on being too busy giving a "how to become a real bitch" lesson as a courtesy.

Therefore, what difference does it make to learn something if one is unable to change their habits? Moreover, when emotions take over, rational thought and resolutions vanish. How often had I lost the opportunity to learn something new or implement what I had learned?

***

Utilizing a similar perspective to the pandemic, what have we learned as individuals? Have you seen any considerable difference in behavior or change of habits in your community since it began?

In Miami, I have noticed that people consume more disinfectants and spend more money on hand sanitizer. They are also investing in paper towels and hoarding toilet paper. But other than that, and masks hiding unbrushed teeth, have I seen any meaningful changes?

Unfortunately, the answer for me is "no." During the

quarantine and the months that followed, I had hoped the pandemic was an opportunity for people to take the extra time while trapped inside their homes to think and contemplate life.

I wished that such an opportunity would allow us, as a community, to come together and learn how to be more courteous to others. To not be always in a hurry. To drive more defensively, primarily given that, for a while, there was no place to go.

I, for once, learned a lesson I am taking to heart. During months spent mainly at the house, I realized I needed to be more productive and manage my time better to no longer be in such a hurry. This applied equally to any personal or professional task I performed, including driving.

As things began to go back to normal, I noticed the difference in my quality of life due to the changes in habits I had implemented. Nevertheless, as the volume in traffic around the city increased, no changes were perceptible overall. Drivers still seem to be in a hurry, as if the streets belong to them and everyone else must get out of their way. Honking and a lack of patience seemed to increase; before the light has been green for a split second, someone in the back of the lane is already punching the horn mercilessly. Zombie drivers, hypnotized by their smartphones, are also back on the streets, so distracted that they almost hit pedestrians.

Ever since we passed the worst of the pandemic, my running mates and I have almost gotten hit by distracted drivers at about a rate of 50 percent more than before. Worst of all, instead of apologizing for the distraction, some still blame us for not watching out. That's even considering we are the ones calling them out to not run over us as they head out of parking lots or prepare to turn right when the pedestrian sign is on.

"Watch out!" an older gentleman yelled, sticking his head out the window as I tried to cross the street while the pedestrian sign began its ten-second countdown.

***

During the pandemic, I realized that I am mostly living up to my personal expectations by enjoying every day to the fullest – as if it is the only one I have. I attribute that partially to practicing one of my favorite hobbies: running. More than exercise, running lets me take advantage of the healing power of the outdoors.

I started with daily walks, which quickly transitioned into daily runs. Very soon, I was running my first half marathon. Along with it came a sense of discipline and commitment. Above all, I learned how strong and capable I am. Running gave me the confidence I hadn't been able to acquire my entire life, even though the challenges I had faced early on taught me to be emotionally strong.

Friends and relatives who are not adept at physical activity may not understand my desire to wake up early to run. However, once we grasp the profound meaning of our lives on this incredible planet – even though we might be microscopic in comparison to the grandiosity of the universe – it makes sense to desire to be out there. Running is not simply for the body and mind; it's also about enjoying life and appreciating the blessings of doing what we do.

Running also offers a wonderful opportunity to socialize. Every Saturday morning after our runs, a breakfast follows, allowing us plenty of time for storytelling, laughs, and toasts, cleansing all the stress from our troubles brought by life or our routines.

I wish all individuals could get into the habit of being active. While I hated to run when I was in high school, I later realized there must be something out there to bring a sense of joy and accomplishment to all of us, even if running is not for everyone. I believe that taking care of our bodies and minds brings tremendous advantages if another pandemic returns in our lifetime.

Not being able to meet with our friends for runs during the first weeks of the lockdown placed a heavy weight on my shoulders, a reason why a few of my running mates decided to still meet, maintaining a larger social distance of ten feet. Having the possibility of being together taken away brought us closer together, even if we missed our good morning hugs.

Eventually, we could no longer take the absence of touch, so we began to touch sneakers as a greeting. Of course, now that most of us have received both doses of the vaccine, hugs and kisses have once again become a part of our encounters.

***

During a flight to New York in December 2020, I sensed a few stares from other passengers; I seemed to be the only one using the distributed sanitizing wipes to clean my family's seats and trays. Even though I am used to being judged and am OK with it, what doesn't please me is seeing that people haven't changed.

After the losses, the grief, and the challenges we have faced during the pandemic, I find it unbelievable that people still are bothered by insignificant actions that do not personally affect them. After having lived through a year filled with opportunities to learn and appreciate deeper meanings of what life is about, it seems many people learned nothing.

As I wiped the seats on the plane and noticed a few stares from others, I wondered if we would walk out survivors of a pandemic but still be that shallow.

Their stares reminded me of the lady at the grocery store who believed me to be sick for taking precautions. Yet, after enduring a pandemic, nothing seemed to have changed. Not that people still don't care to sanitize surfaces, but by the intrinsic need of people to observe and consistently judge others.

I will not be changing my ways. People can stare all they want; to me is a matter of being cautious. And in case you are wondering, being like this is not stressful to me at all. It might be to the people who must live with me, but this is simply the way my brain works: if sanitizing wipes are available, not using them corresponds to not accepting to drink the glass of water set on the table by the server, even if that means no one asked for it. Simple as that.

***

I am a firm believer in common sense. After witnessing the challenges my father faced when losing his battle with cancer when I was 21, and after becoming a mother a decade later, I began to take my health seriously.

I know my daughter will mirror my attitudes; I want her to understand the importance of being healthy, having self-esteem, and knowing how feeling good about ourselves can help us become better and happier. However, the main reason is I don't want her to spend her teenage years like I did, waking up wondering if that was the day my dad was going to die.

It is a sad thought, but that was my reality for an extensive period of time; the factors life had presented to me. Nevertheless, instead of wasting my energy trying to find something else to blame, which would keep me stuck on the problem, I learned to look past it.

The best way to deal with a problem is to dedicate time to contemplating a solution. It could be a day-to-day solution or long-term. Whichever works. Regardless, I believe in using the expertise I have acquired while trying to survive, learn, and grow from specific experiences. Then applying that knowledge in my life to transform lectures into actions.

***

While I recognize what happened to my father wasn't his fault, I know that if I can do my part to become a healthier individual, my efforts will work in my favor if I ever become ill. Choosing a healthier lifestyle enabled me to continue to indulge in special foods and taught me that exercising made me happy. Physical health does help us overall, but when it comes to mental health as part of the pandemic's aftermath, there's a general concern about the consequences of the distress imposed on us in 2020.

Nonetheless, while we seek to understand how we can be prepared if we face another pandemic, it is undeniable that we intrinsically outsource the responsibility, assuming that the government, biopharmaceutical companies, doctors, scientists, and so on are the ones who must come up with answers. Not us ordinary citizens. As a result, we fail to realize that the most important thing is preparing ourselves to withstand such context. How many have lost it during the pandemic?

Consider what happened in my neighborhood. It was late evening in May 2020 when I became startled by unceasing police sirens going by a main street nearby. Later that night, the word was that the police were searching for suspects who had kidnapped a child in the parking lot of a hardware store. In despair, I held on tight to my daughter that night, wondering what torment the poor mother whose son was taken away from her was going through.

The following morning, my phone quickly became mobbed by messages after a friend who had gone to ride her bike around the same golf course my friends and I ran requested us not to use the route that morning. There was a police presence, and the streets were blocked. The area had become a crime scene.

A body had been found in shallow water on the golf course. All concerns were directed toward the child kidnapped the night before. A cyclist noticed a body floating and contacted the police, who confirmed the body was of the boy who had allegedly been abducted.

Nevertheless, it didn't take long for the investigators to learn that the mother was the one who had murdered her autistic son by pushing the boy, who couldn't swim, into the water. After two or three attempts in different locations, she was finally able to accomplish her goal at the golf course.

Given the quarantine and the impossibility of finding caregivers to watch her severely autistic child, she no longer could take the stress. "He is in a better place now," she allegedly told investigators.

For months, I couldn't run around the golf course without nearly vomiting when I passed the memorial filled with stuffed animals and balloons. Even if I looked away, the sense of dread was still there. However, I remembered that I was blessed for surviving a time of madness and promised myself I would not stop running since it was helping me stay sane during challenging times. I'm glad I didn't stop; later that same year, a family member in Brazil committed suicide.

***

During a recent read of the book *Endurance* by Alfred Lansing, first published in 1955, a specific passage caught my attention. It's where the author shares the experience of Shackleton's expedition to the Antarctic along with twenty-seven crew members onboard a ship. They were stranded in the Weddell Sea during the Antarctic night. They could not see the sun for seventy-nine days while stuck in ice floes that would not break apart. However, after being at sea for nine months, the men's closeness helped them fight depression, the opposite of M. le Baron de Gerlache's experience.

In 1897, de Gerlache was the leader of an expedition to the Antarctic and contributed to the construction of the vessel that Shackleton would later rename *Endurance*. According to the author, "by coincidence, the man who had once been a partner in *Endurance* [the ship] had himself been beset in the Weddell Sea aboard a vessel called the *Belgica* in 1899. With the coming of the night, the *Belgica's* crew became infected with a strange melancholy. This slowly deepened into depression and then despair as the weeks went by. In time, they found it almost impossible to concentrate or even to eat. To offset the terrifying symptoms of insanity they saw in themselves, they walked in a circle around the ship. The route came to be known as the 'madhouse promenade.' One man died of a heart ailment; another gave way to hysteria."

No matter how difficult the circumstances in our lives, they are not impossible to overcome. Furthermore, turbulence is not what defines us but how we face it, perceive such experiences, and rise above them. Shackleton is an excellent example of this.

Running represented a way out of the madness for me during the pandemic. It kept me from sinking into the tragedy the world had become. I figured out what works for me, giving me the confidence to say I am prepared for the future if another pandemic strikes.

I am also sure that I will write if I can't run. And if I can't write, I will surely find something else to keep me sane. I will scribble on the walls of my entire house before I allow myself to be overtaken by madness. And if I no longer have walls, I will find animal shapes in the clouds. And if there are no clouds outside...

It's a matter of mindset.

***

As I watched other parents weep over the fact that Disney World would never be the same, I focused on the extraordinary amount of time the pandemic had given me to spend with my daughter, trying to make the best out of it despite the challenges of virtual learning.

Although my six-year-old well understood the concerns over the coronavirus, not once did she see me crying, worried, or depressed about the pandemic. That doesn't mean I was completely alienated from it, but I believed it was my responsibility as a parent to create a safe environment. She, indeed, felt that sense of safety.

Unless we learn to shift gears toward mindfulness and embrace the idea of a healthier mindset, I do not believe we will ever be prepared. Hence, our children who tend to learn by example – and who will be the citizens, doctors, scientists, leaders, etc., in a future pandemic – won't be prepared either.

My daughter missed her friends and became sad if she couldn't have a playdate with them, but she also understood that being together wasn't an option. At least for the time being. She comprehended there were too many risks involved. Therefore, she found ways to feel closer to them: when she wasn't making drawings and cards to send to them, she would FaceTime with them, allowing closeness even at a distance.

Ballet classes also went from in-person to Zoom. It wasn't the same experience, but it helped. The schools found solutions, the ballet teacher found solutions, and so did my daughter. Not problems: S-O-L-U-T-I-O-N-S.

***

I can't imagine undergoing a pandemic without the comforts and technology we currently have available. There are many things for which to be grateful. Virtual school comes to mind. We wonder what the future holds for the students who had their education affected during remote learning. Such inquiry makes me turn toward the past and wonder what happened to the children who experienced the Spanish Flu pandemic one hundred years ago.

While I think it would appear more relevant to come up with predictions based on profound data analysis as to the consequences for students, I would like to state that the implications of remote learning are not a matter of hopelessness. In other words, it's reversible.

In the book *Team of Rivals,* author Doris Kearns Goodwin shares the story of our sixteenth president,

Abraham Lincoln, and his three rivals during the nomination of the Republican Party candidate for the presidency in 1860. In the book, she considers their backgrounds. Henry Seward was selected among his brothers for a college education. "His destined preference," as he called it, led him to a preparatory academy at the age of nine. His day began "at five in the morning and closed at nine at night." Like Seward, Salmon Chase was also picked out to "receive a better education." According to the author, Edward Bates "was taught to read and write by his father and showed a talent for poetry by the age of eight."

For his part, the author notes, Abraham Lincoln "faced obstacles unimaginable to the other candidates for the Republican nomination. In sharp contrast to the comfortable lifestyle the Seward family enjoyed, and the secure early childhood of Chase and Bates before their fathers died, Lincoln's road to success was longer, more tortuous, and far less likely."

I have visited the Kentucky birthplace of Abraham Lincoln. It's still a small and simple log cabin sitting on the outskirts of town, surrounded by a green field.

Nonetheless, as the author corroborates, "In later life, Lincoln neither romanticized nor sentimentalized the difficult circumstances of his childhood. Allowed to attend school only 'by littles' between stints of farm work, the aggregate of all his schooling, Lincoln admitted years later, 'did not amount to one year.' He had never even set foot 'inside a college or academy building' until he acquired his license to practice law. What he had in the way of education, he lamented, he had to pick up on his own." She concludes, "Books became his academy, his college."

With Abraham Lincoln's record in mind, I would say that any child born a citizen of the United States who was affected by remote learning, like Abraham Lincoln, still has a chance to become president with dedication and hard work. Looking into past events of great significance might steer us in the right direction.

***

Because of the Great War, the global economy declined. Additionally, the Spanish Flu pandemic began in early 1918. Despite the effects on the economy, post-war advances in technology were substantial, allowing social change, such as communicating and traveling.

During World War II, women were hired for jobs previously designated only for men. As a result, the economy presented what is known as the Golden Age of Capitalism – a period of sizeable economic expansion on a global scale.

Just as these post-crisis periods presented social and economic changes, COVID-19 has undeniably brought the advent of behavioral patterns, such as using grocery shopping services or, overall, making purchases online.

By way of illustration, an online retail store based in South Florida had sales increase by 47 percent in 2020 over the previous year. In addition, on Cyber Monday 2020, this retailer acquired 50,000 new customers – 85 percent more than average. At the same time, I am sure that some businesses will not be able to recover, partly due to being unable to bring any revenue during the shutdown, while their financial responsibilities remain.

In my view, a recovery will also depend on the various business markets. If we base economic revival solely on businesses directly affected by the pandemic, such as cruise lines, I think full recovery will take a great deal of time. But I believe it's relevant to analyze other sectors that are prospering post-pandemic.

Relocation is also a great possibility. I know an individual who, after working several years in the cruise line industry, received notification of an impending layoff. Prior to the pay cut, she managed to find another job occupying a similar position in another field.

Also, the online business I used as an example above

never ceased hiring, even during the lockdown. After being unable to find local workers, it went as far as seeking out-of-state candidates. The company is currently expanding its operations and is opening new offices in other states.

Furthermore, the real estate market in South Florida is currently strong. Inventory is low, home prices continue to climb, and buyer activity remains high. For example, in February 2021, a homeowner could sell in twenty-four hours. While home prices have increased 7.8 percent since last year, Zillow predicts they will rise another 9.4 percent in the next twelve months. According to Norada Real Estate Investments, "In June, a report released by the Miami Association of Realtors said that they expect an increase in sales in the coming months due to falling mortgage interest rates. They're also seeing an influx of refugees from high-tax states. For example, thousands of high-income earners are relocating from New York to Florida. The realtor report stated that there was a 6.1 percent annual growth in overall sales."

As far as Florida's commercial real estate is concerned, according to MMG Equity Partners, the state has been considered a top destination for San Francisco Bay area tech firms in search of new homes.

Despite the exciting news for Miami and South Florida's economy, I believe a national and global recovery of the economy will also depend on the measures taken by our leaders. The product, once again, is a result of the multiplication of the factors.

***

Uncertain of what the future holds, I conclude by saying that while I am in no position to control anyone's minds or how others behave, I can make a personal statement concerning the aftermath of COVID-19.

My personal commitment is to take advantage of this crisis and continue implementing the changes necessary to evolve as a human being. I also pledge to focus on my personal growth to be a better individual.

Also, I plan to continue helping others to overcome challenging times. I believe we ought to become mentally stronger, to be less victims of circumstances. We shall learn how to be creators of our own destinies.

In all honesty, my only expectation for life after COVID-19 is to see people smiling and laughing more – even if that means laughing at ourselves.

I am confident that as a society, in multiplying

efforts to become happier individuals and consequently better humans, no number of digits will be capable of reversing the product's state of pricelessness.

Together, even when apart, we are stronger.

About the Author

Nanda Klein is a creative nonfiction author. Born in Brazil, she has lived in Miami, FL for the past several years. In her personal blog, The Nanda Klein Journal, she shares personal experiences and talks about her life journey, turning challenging times into lessons learned. She describes her blog as being "about the good and bad in life." To learn more about her, please visit www.nandaklein.com.

astarot / Shutterstock.com

Outdoor exercise was one of the few activities still allowed during the COVID-19 pandemic. Some people felt it necessary to wear face masks in this setting. Something must be working because this couple seems to be flying!

AFTERMATH

"None of us, including me, ever do great things. But we can all do small things, with great love, and together we can do something wonderful."

Mother Teresa

Fool Me Twice

Mark Lyons

Shame On You

"Good morning. How are you today?"

"I'm very grateful to be getting my shot."

"Did you have to wait long?"

"Just a few minutes, but I've been trying to get an appointment online for three weeks."

"I know you just filled out this form, but I want to ask you some of the questions one more time because they are so important. In the last ten days, have you had a cough, sore throat, chills, or fever?"

"No."

"Have you ever had a severe reaction to any vaccination or medication?"

"Yes, I am allergic to penicillin."

"What reaction did you have after taking that?"

"I had a rash all over my arms and chest."

"We are looking for any life-threatening reaction, like difficulty breathing."

"No."

"Have you had any vaccination of any kind in the last fourteen days?"

"No."

"Have you ever tested positive for COVID-19?"

"The question said in the last ten days."

"That is true, but we are also concerned that you may have had monoclonal antibodies, and I know that answer would be no if you never had COVID-19."

"No."

"Do you carry an EpiPen for any reason?"

"No."

"Which arm would you like to use?"

"My left. That didn't hurt at all. I barely felt it go in."

"Small needle, small dose, a good combination."

"You are really good at this."

"I've been practicing."

***

Barely five years since the last scourge on human life was effectively corralled, we again face a new challenge to our lives and sanity. They are calling this one COVID-28, as it is another variant of the one that ravaged the world the last time and was dubbed COVID-19. From what I heard on the TV, this will make COVID-19 seem like a cakewalk. They say that this virus, unlike the other known variants, is easier to transmit to other people. The masks used so often during the first COVID pandemic are essentially useless during this one. Even the better N-95 masks that were so hard to breathe through, and therefore thought to be more effective, are only slightly better than nothing this time. It takes only a tiny fraction of the new virus to cause infection.

I remember hearing that one of the first symptoms

of COVID-19 that so many people detected was a loss of smell and taste. The report said that the first symptom of COVID-28 is a sneezing fit that can go on for several minutes. I now remember finding humor in the stories of ladies in the early stages of the first viral disease, not realizing they were infected until one of their lady friends would tell them quietly that they used way too much perfume. I guess a long spell of sneezing will be less subtle, though likely more deadly to those who resist using the newest virus protective gear.

When COVID-19 was just getting a hold on the world back in 2020, I worked in an assisted living facility for elderly and mildly demented people, most in their eighties and nineties. We thought we were doing well with our program of checking the body temperature of everyone who entered the building, and for a while, the governor of the state banned visitors in places like ours. Most of our staff had no medical credentials, not even certified nursing assistant (CNA) diplomas. They were all excellent caregivers, but with no real training in infection control, I feared that any of them might unknowingly spread the virus throughout the building.

Everyone was issued masks and given brief instructions about handwashing and the use of rubber gloves when engaged in direct patient care. Initially, the water at the sinks that seemed to take forever to warm up sufficiently was always warm. Everybody was worried enough to take the most basic precautions. My bosses took pride in the fact that we had no cases throughout the summer and well into the fall. We could all see the stress of the situation beginning to show on the face of our administrator, Rick, and he frequently told us that he could no longer sleep at night. He would regularly thank us for our success in keeping the virus from our facility but always followed that with a summary of his discomfort and fatigue. Each time the staff left one of his town hall meetings, as he called them, we would smirk at each other and then refer to the meeting as an "Oh Poor Rickey" session.

***

"Good morning. Please take a seat. You brought your husband for support. That is a great idea if you are nervous."

"I'm not afraid of the shot. The first dose was nothing, but I faint easily, and I have fainted after injections before."

"Recently? How long has this been occurring?"

"For years. Now it has started happening to my daughter."

"Did you faint after the first dose?"

"No, and I have a letter from my doctor saying I should get it. Here, you can read it."

"This is a Medically Vulnerable form that does not specify any particular condition, but I can honor his request and give you the shot. You can lean against the table, and your husband can stand on the other side. What does your neurologist say about these spells?"

"I have never gone to a neurologist. I always know when it will happen."

"How do you know?"

"I hear a ringing in my ears, and my visual field shrinks. When it gets really small, I go to sleep."

"After you wake up, are you tired?"

"Yes, my daughter and I get sleepy after we faint."

"Your doctor never referred you to a neurologist?"

"He said I would grow out of it."

"I strongly suggest that you consult one; he may be

able to help you avoid these spells."

***

I gave the woman her shot as she and her physician asked; she did not faint. Her fainting symptoms seemed consistent with a vasovagal fainting response to pain or stress, but her sleepiness after the episode made me think a seizure disorder could also be possible. It was satisfying to direct her to a neurologist, and it was fun for me in that most repetitive job.

My confidence in effective infection control practices dates to the mid-eighties, during the first pandemic of my nursing career. AIDs killed everyone who contracted it, and I worked in a nursing home populated by its victims. Initially, we were as scared then as the uneducated staff at the assisted living facility (ALF) 35 years later. There was much uncertainty back then, and our daily infection control meetings evolved into teaching theory as much as technique.

It wasn't difficult to teach staff to wash hands properly or never contact a patient directly without proper personal protection. However, we knew that we couldn't prepare them for every situation they might encounter. We didn't realize it at the time, but AIDs, a bodily fluid-borne virus (e.g., blood), was a much smaller risk to staff than a virus like SARS-COV-2 that could be spread through the air or from touch.

The practice of universal precautions emerged during those days, and we were taught to treat every patient the same, with an AIDs diagnosis or not. We even discussed, but later discarded, the idea of not telling the staff which patients had AIDs and which did not. In theory, it should not have mattered, but we weren't living a theory, and each staff member may not have understood every aspect of the plan. Even for highly trained nurses and CNAs, there remained fear of a breakdown in technique over the long run.

***

My work schedule that autumn at the ALF was part-time, two or three days a week. Most days, there would be an exposure to virus spread that I could recognize, and I would point it out to my nurse supervisor or Rick. Usually, my suggestions were met with nods only. As far as I know, none were ever acted upon.

I pointed out that nurses from outside agencies frequently came to our facility after working at known COVID-19 receiving facilities the previous shift, never changing their uniforms nor providing recent virus-free test results. If they had no fever, they qualified for the shift. A couple of times, one of our staff members came to work, obviously ill but deemed safe by the receptionist who checked their temperatures at the front door. Both times I sent that person to the nurse supervisor for further evaluation. One was sent home, and the other returned to the floor. There was no question in my mind that the level of respect for the virus was eroding, and I recorded several issues to bring up at the next town hall meeting.

There were no other staff nurses at that late November meeting. Social distancing was not enforced. The administrator, nurse supervisor, ten or so caregivers, and I listened as Rick read aloud the horrible case reports and death tolls from the most affected facilities in the county. He referred several times to the stress he felt and his ongoing insomnia. The nurse supervisor said nothing, deferring to Rick on all issues faced by this group. He spoke only about using hand sanitizer, with a maximum of three consecutive times between hand washings.

As soon as I got an opening to read my list of concerns, I recited everything I reported previously, one unanswered issue at a time. Rick appeared irritated that I brought up any of the things I thought were important and did not comment on them.

Before he could shut me up and return to his personal story of COVID-19 fatigue, I made it a point to tell the group that I was monitoring the water temperature at our main hand washing sink. The water had always been warm when we began taking precautions; now, it was always cold. Still no response.

Rick ended the meeting with a telling statement. He said that during a conversation with the county health department, they said it was inevitable that there would be cases in our facility. You could hear his relief after relating that conversation. His personal responsibility would be eliminated if the health department said there was nothing he could do to stop it.

***

I had a few days off after that meeting, and during every waking minute I was consumed with the idea that I should resign from the job. Those people didn't care about anything I said, and their efforts to prevent catastrophe were waning. The decision was not coming easily to me. The twenty-four residents on the unit I usually worked were all dear to me, and a sick feeling gripped my stomach when I imagined any of them getting sick from the disease.

Friday was my next shift, and that would be the target day for providing my notice if I decided to leave. On Thursday evening, at seven, I was still undecided about resignation. Then, I stood up from my chair and felt it for the first time. With no earlier symptoms, I was now woozy, and an intense chill crept up my back. By the time I made it to my bed, I was shivering severely. There was no question in my mind about what was going on, but I held out a small amount of hope that I was wrong. The following day, with the aid of acetaminophen, I felt a little better. Still, with a low-grade fever remaining, I called work to tell them I would not be coming in. By mid-morning, I sat in my car in the parking lot of the clinic that would give me a rapid antigen test. Finally, after several hours, I was summoned to the testing room inside and was told that my test was negative. I was never as happy to be wrong about anything that I could remember and rationalized that I may have an optimistic diagnosis of influenza.

My good sense returned when I realized that I had never felt that way when it was the flu. On Sunday, both my rapid test and polymerase chain reaction (PCR) tests were positive, and I officially had COVID-19. Twelve days later, my fever broke, and for a short time, I felt comparatively better. Then pneumonia set in.

***

I returned to work on a minimal schedule on Christmas Day, only to find that my floor of twenty-four was now eleven. Everyone on the floor tested positive, and only the efforts of our great physician saved the surviving eleven. I was shocked that our previously successful efforts to stop the virus from spreading were ineffective for even one of my dear residents.

After the horrible news of the destruction of my unit at the ALF, I was angry at the bosses for their inability to prevent the transmission of the contagious disease. However, it only felt right that I should participate in something to help save lives, so I signed on with the State of Florida to administer the new vaccines, including the new messenger RNA vaccines from Pfizer and Moderna.

Though I had been a nurse for over forty years, this was the first time that I genuinely felt a calling. I hoped that the natural immunity I derived from my infection would be enough to protect me from a second round of the disease until I could get my own two-shot vaccine. It worked. For several months I administered about 125 shots a day. Even though the long shifts were tiring for someone of my age and lingering respiratory symptoms, the effort was the most satisfying of my career.

***

"Good morning. How are you today?"

"Fine, just glad to get this done."

"I'd like to run through these one more time because they are so important. In the last ten days, have you had any fever, cough, chills, or shortness of breath?"

"No."

"Have you ever had a severe reaction to any medication or vaccination?"

"No."

Have you ever tested positive for COVID-19?"

"I had it in late December. Both my wife and I got sick at the same time."

"Did you receive monoclonal antibodies at that time? Unfortunately, we cannot vaccinate you if you received them in the last ninety days."

"I did not receive them, but my wife did. My symptoms were mild, but hers were very bad. I knew I was sick, but she had a high fever and was very lethargic. Three days later, I convinced the doctor to put her in the hospital, and three days after that, she died."

"I'm so sorry for you and your family."

"Thanks."

"Which arm would you like to use today?"

"I don't care."

***

When I was in high school, Professor Paul Ehrlich of Stanford University published a book called *The Population Bomb*. It detailed his predictions about the over-population of the world and the consequences involved. At the time, his dire predictions, though frightening, seemed a long way off and not something that would be a significant issue in my lifetime. Yet, for fifty years, the warnings that book posited have stuck with me as each potential human coffin nail is forged. Whether any specific prediction is on target or not, the grand message was clear and remains so today. The world population has doubled since the book was published, and the warnings about famine and climate change due to overcrowding are still relevant today. As the superior species on this planet, we seem unable, and in some cases unwilling, to do anything about it.

Lower life forms seem able to avoid extinction, or at least take steps instinctively to that end. A new hill of fire ants in my yard will scramble around and lose many members when I dump a bag of poison on them. Then, they will show up again in a day or two, appearing healthy and with no apparent memory of the attempted insecticide. When the destruction of habitat reduces the size of the range of coyotes and bobcats in Florida, they adapt as well as they can, often to the detriment of domestic chickens. All lower species have the instinct of survival, and they use it predictably. It seems very strange that the one species that can understand and reason cannot do anything about the steady march to their own destruction.

We can easily see that the climate of this planet is deteriorating. We can see food and water shortages in densely populated areas. Scientists make frequent references to past epidemics like they are a road map for our behavior that will prevent the worst outcome. They advise us to use the precautions that may be our only steps to safety. Some of us take those admonitions as sound advice, but others stubbornly reject everything as if there are other motives involved.

During the 2020 COVID-19 problem, I would be dressed like an astronaut at the local grocery. Others, with a different attitude, did nothing obvious to prevent their own infection or prevent the asymptomatic spread of their possible diseases to others. I often heard people say they had the right to go without a mask, which to me indicated their indifference to the risk for everyone they met. Stores that proudly advertised their mask-wearing policies declared their inability (or desire) to enforce that policy.

By the latter months of 2021, when unvaccinated people began dying quickly in large numbers from new variants, many non-mask wearers finally began to heed the warnings. Unfortunately, it was too late and no longer as effective for many. Some of the more aggressive mutations seemed to penetrate the masks and infect people who might have been protected from the early strains.

For those of us who used every precaution we could think of, the choice to stay away from anyone else was the only sure preventive measure. Everything may have turned out differently, and I mean better, if the early deniers had taken the precautions I took. Herd immunity may have changed the balance between virus spread and virus suppression. Many of the virus' mutations may have never evolved because they only occur when the viral DNA replicates in the victim's cells. Many lives could have been saved, and many people with long-term COVID-19 damage to their lungs and other organs could have been spared the chronic effects.

***

Fire ants and bobcats instinctively react to

encroachment into their territories, but humans have so many other influences to cloud their thinking and decision making. A nation divided by political issues creates a polarization of the population on all issues. Suppose someone would normally act reasonably about virus spread precautions. In that case, the political pressure applied by a leader who does not believe such things or who has a motivation that is not known can be behavior-altering. If that beloved leader says something or dismisses rational thinking, the followers will often rationalize their behaviors to match them. They purposefully deny the obvious to declare their loyalty to their chosen leader. Add the reinforcement of their like-minded friends, and the battle lines are drawn. They may feel strongly about personal rights or race relations, and then believe all their leaders tell them on various issues without examining the facts.

In March and April of 2021, when a vaccination patient came to my table after refusing to be serviced by a black colleague, I felt some anger, as I had witnessed discrimination against other black nurses before, both in comments and actions. In my group of approximately 100 nurses, over 75 percent were black, and over 90 percent were women. While many of the group had only a small fraction of my nursing experience, I was always comfortable working around black nurses and respected their devotion to their patients. When this white woman came to my table, I wrongly assumed she was racist. Without my encouragement, she just opened up about her choice to find another vaccinator.

***

"You know that nurse has not taken the vaccine?"

"I know that many of the nurses here have not. Is that why you came to me?"

"Yes. Aren't you afraid they will catch it from one of these patients lined up around the corner and down the street? None of them are vaccinated. The nurse could pick it up from an infected patient and give it to another. Are you vaccinated?"

"I have been fully vaccinated since February eighteenth. It has occurred to me that there are risks, but everyone takes all the precautions available."

"I was surprised that the nurse was not vaccinated. Isn't it mandatory for you to get the shots before giving them to the public?"

***

That patient's fear stuck with me that spring, and I decided to conduct an unofficial poll of the nurses who would share their vaccination status. Most were willing to tell me if they had been vaccinated, but several who opted not to get the shot refused to explain why they decided to forego the protection. One black nurse spoke about the horrible tests conducted during the Tuskegee Syphilis study of 1932, where 600 black men were involved in a study about treatment of the disease without informed consent. I suggested that 90 years had provided us with many changes that will prevent a recurrence of that type of injustice. She disagreed.

Before my poll, I was aware of only two black nurses that were fully vaccinated. Much of the information I gathered came from one member of the small teams, which made up our one larger team. My usual approach was to ask a white member if she knew the status of the other nurses on their team. Most were fully aware. Florida required all nurses in the program to have weekly COVID-19 tests unless fully vaccinated. When the testing time was announced each week, nurses were required to respond on the communication app if they were free and planned to come to the testing site, or if they were at a vaccination site and were unable to attend at that time. By keeping track of those people for two weeks, I determined the vaccination status of 69 of the 100 plus nurses in our group. Of that number, I found that only twenty were vaccinated. Of those twenty, I knew that sixteen were white. While the exact percentages were unnecessary for my informal poll, it became clear that a considerable imbalance existed between black and white nurses.

Our team held several weekend pop-up vaccination clinics at black churches in Jacksonville. Those who flooded in to receive their shots were no different from the general populations we served during the week. Everyone was friendly, and in most cases, very happy to get vaccinated. The only difference I could detect was age. Those at the church sites were generally older and more afraid of the virus.

The nurses were all much younger and, from my perspective, easily led by their co-workers. I wondered if the freedom to choose caused the black nurses to exert and demonstrate their ability to decide for themselves, as a statement of independence. With all the numbers swirling in my head, I decided to try to find an answer to the question posed by the lady patient who refused the vaccine from an un-vaccinated nurse. A call to the state COVID-19 hotline brought some clarity, but not the comforting kind.

***

"Hi, I just got my vaccination at one of the state sites. The nurse told me that she had not received the vaccine herself. Is it true that the vaccinators are not required to be vaccinated before giving shots to the general public?"

"Everyone has the right to choose," the operator replied. "We can't force anyone to get the shots."

"I'm not so sure that is true. I've heard of many hospitals and other health care facilities that have made it a policy for all staff to be vaccinated as a requirement of employment. And many colleges are planning to make full vaccination a requirement of admission."

"It is the employee's right to choose."

I was aware that some businesses and healthcare facilities required vaccination as a job requirement, but the Florida state governor had consistently demonstrated economic goals over health safety and had banned any form of vaccine passport requirements. No one could be required to carry vaccination records to enter a place of business. Many companies claimed they would suffer losses from such an approach, but no one spoke about such a requirement for all people being an equalizer for all businesses. The ban seemed to fit well with the political positions held by the governor's party. It was reported that 47% of his party did not plan to get the vaccination.

***

Even with almost sixty years since Dr. Erlich sounded the alarm, we humans have shown very few signs that we are heeding the warnings, nor even demonstrating the desire to change the outcomes.

There is one comforting sign that the Earth is capable of healing itself. Several pollution and global warming indicators improved during the first COVID-19 crisis in 2020. For example, people used their internal combustion engines less by working from home and not traveling as much overall. This development begs the question: How many people can the planet sustain without causing the global warming, severe climate change, and pandemics that will commit most of the genocide necessary to return the Earth and its inhabitants to a static existence? Can the human population self-regulate enough to keep deterioration from occurring further?

Population numbers are doubling increasingly rapidly. It seems doubtful that anything short of a catastrophe can save the planet. A person in a third-world country who spends every waking hour trying to feed themself to survive can hardly be expected to worry about the world's survival. Of course, that person probably does not contribute a great deal to global warming. An American with a gas guzzler and expensive tastes contributes much more to our destruction than that man who only tries to feed himself.

***

Shame On Me

"Good afternoon. Please take a seat."

"Thanks. I am happy about getting this over with."

"Have you been waiting long?"

"No, you people are well organized."

"Thanks, but my work is very task-oriented. However, the county is very good at this kind of event."

"I would like to go through these questions one more time just because they are so important. In the last ten days, have you had any chills, fever, cough, or shortness of breath?"

"No."

"Have you ever had a severe reaction to any medication or vaccination?"

"No."

"Have you ever tested positive for COVID-19?"

"My husband and I both had it twice. Once in August and again in December."

"You are the first person I have met in all of these weeks that had it twice."

"We were very sick both times. I wouldn't be surprised if anyone who got sick twice did not survive."

"Did you have monoclonal antibody treatments?"

"Yes, right before Christmas."

"I'm sorry, but that was only 75 days ago; I cannot give you the vaccination until 90 days have elapsed."

"Really? That's ridiculous."

"I am sorry, but that is the CDC guideline."

***

It took more than four years to stop the worldwide deaths from COVID-19. Millions of people suffered and died because of the pandemic. There were not enough deaths to diffuse Erlich's Population Bomb, but there were enough to wake up many people to the needs of the Earth and the people who live so closely together. Though not tightly related to the viral pandemic, there became much greater interest in all things related to our sharing of the planet with other people from other cultures. People began to characterize global deterioration as a problem we all share instead of a problem caused by others. Countries began earnestly attempting interventions to stem the tide of carbon flooding by working together instead of blaming each other. Carbon fuels that the rich hoped to keep for gluttonous use were in very short supply and very costly. The Earth was beginning to show gratitude for humankind's efforts. The ozone layer hole began to close slightly but measurably in 2020. The once smog-shrouded cities in America, India, and China now have cleaner air. The number of wildfires, hurricanes, and freak windstorms has been at century low numbers in the last three years.

With a new pandemic at our doorstep, those who suffered during the last one in 2020 wonder if we can survive another. Are the lessons we were taught the last time fresh enough in our memories to make us more cautious earlier in this scourge. Can we keep it from spreading before this version takes millions more lives? History suggests that the answer is no. Unfortunately, humans seem to remember the past, but only for conversation. The human race doesn't seem to be able to use history to its advantage.

***

"Good morning. Please take a seat. Have you been waiting long?"

"Not long at all."

"Today is your second dose of the Pfizer vaccine. How did you do after the first?"

"Only a little soreness in my arm. My wife and I drove all the way here from Miami again so I could get this shot. She is 49, and they say she is too young until the governor lifts the restrictions. Is there any way she can get her first dose today?"

"You could try one of the federal sites, but this is a state site, and we are forbidden to vaccinate anyone who does not meet eligibility requirements."

"That is stupid. We live together."

***

During the vaccination missions, many people expressed concern about the mRNA serum attacking their DNA. My explanations, though not professional, centered around the vaccine preventing the virus from entering cells to reproduce its DNA. I explained that mutations are formed through the infiltration of the cell and replication there. Variants are significantly reduced using the vaccines, but the mutations occur more readily when the virus remains in a person's body for a long time. Those getting the vaccine usually took comfort in those statements, but some demonstrated their skepticism non-verbally.

***

"Good morning. How did you do after the first shot?"

"My arm was pretty sore, but I took tomorrow off from work in case it happens again."

"The side effects of the first dose seem to be a pretty good indicator of how it will be after the second, though we don't get much feedback on the second dose."

"Why is that?"

"We don't give a third shot, so no one comes back after the second."

"Of course. My wife said the second one has much

worse effects."

"Is that how it was for her?"

"No, that is just what she said. She won't take any of the shots. She said it isn't necessary. So, if a lot of people do get it, she won't have to."

"Why is she so worried about it?"

"She says that the country's leaders don't think it is important. For a long time, she thought the whole thing was a hoax."

"I can assure you, COVID-19 is real. I was very sick, and my daughter was also ill for a week and a half. Minor side effects are nothing compared to what could happen if you get infected."

"She said she is afraid that she will have a horn grow from her forehead in a couple of years, or some other strange effect that China has sent through this virus."

"The virus may have come from China originally, but the strain that tore through the US came through Great Britain as a mutation from the original. The vaccine can stop the virus in its tracks, but only if people take it."

"I'll try to convince her."

***

Toward the end of 2021, people who were doubters did start to believe the science in large enough numbers that we were eventually able to achieve herd immunity in most of the country. Unfortunately, we trailed most other countries in this milestone because of paranoia fed by political leaders having ulterior motives for discouraging vaccination. In many cases, only when someone dear to them was stricken did people relent and take precautions and get a vaccine.

By the summer of 2022, many more Americans had lost their lives. Scientists now say that few people who died at the end of that period had complete vaccinations. They said that anyone fully vaccinated had nearly zero chance of dying from the original COVID-19 virus or any early variants. That information came out the year before, when there were only five hundred thousand deaths. Each year since early 2023, there have been many tragic deaths even though COVID-19 was declared eliminated from the country.

***

"Good morning. Please take a seat, and I will look over your paperwork. This all looks fine, but I would like to go over some of these questions because they are so important."

"I just filled out the form. My answers haven't changed in five minutes."

"I realize that but the answers you gave are yes or no, and some of the questions may lead to more detailed answers that are important for your safety. For example, in the last ten days, have you had any cough, fever, chills, or shortness of breath?"

"I already answered no, but I've been sweating more than usual today. It is hot already this morning. You are the one coughing. Maybe you should answer these questions before you give shots to a bunch of people."

"Sorry if I coughed in your direction, but I inhaled your perfume, and it tickled my throat."

"My perfume is choking? It's very expensive."

"Did you eat breakfast today?"

"Of course, but knowing I was coming here kind of killed my appetite, so I forced myself to eat tasteless bacon and eggs."

"I would like to check your temperature before giving you your injection."

"Fine. Let's get on with this."

"100.8 temporal. That is the equivalent of 101.8 orally. I'm not comfortable giving you the injection. I think you should get a COVID-19 test first."

"Is there a manager or a supervisor here?"

"Yes, Leah is the site manager."

***

Will the Earth use this second pandemic in nine years to return to the healthy environment our recent ancestors enjoyed by thinning our numbers? If humans can use the knowledge we gained from the last threat, we may also see that there is some hope for the planet's future. If we work together, maybe we can act as one to solve the other vital problems we face.

Can we recognize that these fast-spreading diseases are warnings to us that forces beyond our control will govern our future if we cannot heed the warnings soon? If we can use what we have learned, shun the efforts of those who question the science, and use the previous experience as a guide for dealing with the other outstanding issues we face, there is real hope for this planet. This newest pandemic is a test we must not fail.

About the Author

Mark Lyons has published several mystery/crime novels. He is a long-time nurse and former farmer. He spent much of the time during the COVID-19 pandemic away from home administering the coronavirus vaccine to Florida residents as a state official working through the program formerly called Operation Warp Speed.

PHYSICIAN TO THE PRESIDENT
THE WHITE HOUSE

24 July 2022

MEMORANDUM FOR: KARINE JEAN-PIERRE
ASSISTANT TO THE PRESIDENT AND
WHITE HOUSE PRESS SECRETARY

FROM: KEVIN C. O'CONNOR, D.O., FAAFP
PHYSICIAN TO THE PRESIDENT

SUBJECT: President Biden SARS-CoV-2 Update

President Biden completed his third full day of PAXLOVID last night. His symptoms continue to improve significantly. His predominant symptom now is sore throat. This is most likely a result of lymphoid activation as his body clears the virus, and is thus encouraging. His rhinorrhea, cough and body aches have diminished considerably. His voice remains a bit deep. His pulse, blood pressure, respiratory rate and temperature all remain normal. His oxygen saturation continues to be excellent on room air. His lungs remain clear.

https://www.whitehouse.gov/.../Update-7.24.pdf

According to the Associated Press, President Biden is fully vaccinated. He received two doses of the Pfizer mRNA vaccine before taking office, a first booter shot in September 2021, and a second booster in March 2022.

He still caught COVID-19.

AFTERMATH

Excerpt from press briefing by White House COVID-19 response team and public health officials.

January 26, 2022

DR. FAUCI: Since September of 2020, there have been five SARS-CoV-2 variants of concern: Alpha, Beta, Gamma, Delta, and now the current Omicron. And so obviously, innovative approaches are needed to induce broad and durable protection against coronaviruses that are known and some that are even at this point unknown. Hence, the terminology "pan-coronavirus vaccine."

But I want to explain why this is somewhat of a complicated issue. ... I don't want anyone to think that pan-coronavirus vaccines are literally around the corner in a month or two. It's going to take years to develop in an incremental fashion.

Dr. Anthony Fauci.

https://www.whitehouse.gov/briefing-room/press-briefings/2022/01/26/press-briefing-by-white-house-covid-19-response-team-and-public-health-officials-80/

M.E.E.K.

Richard A. Marschall

It is said the meek shall inherit the earth (Psalm 37:11). Was COVID-19 a biblical quirk, or were our politicians, the deep state, the elite, and the establishment privy to something we couldn't begin to understand?

Two days before the first announcement of COVID-19, the director of the FBI posted an ominous, obscure tweet announcing the formation of a new agency in the FBI's Criminal Justice Information Services (CJIS), to be dubbed "M.E.E.K." The short memo read: "Under Congressional Directive H.R. 89417H, the CJIS has created a new department, to be affiliated and integrated with the National Security Administration (NSA). This joint endeavor blends the elements of human psychology with the newly formulated theories of human cerebellar molecular activity, initially developed at MIT in 2001."

When pressed for additional information later in the day by a reporter who had taken the time to read the memo, the director replied, "I am not at liberty to say anything else at this time. Details will be forthcoming."

This was probably the most newsworthy item related to national security since the NSA revelations a few years ago. Sadly, the media did not press the director for more information. A week later, no one, other than a handful of right-wing members of Congress and conspiracy theorists, even remembered the obscure announcement. The FBI director quickly faded out of the public eye, as did any reference to M.E.E.K.

***

In a confidential conversation between the president and the FBI director the week following the initial announcement, the president asked: "Director, do you think it was wise to even mention to the public the formation of H.R. 89417H? I'm concerned that the congressional committee that proposed this legislation may have ulterior motives for the program's inception. It's been two years since the initiation of the program anyway. We could have left it on the backburner, and no one would have been the wiser."

"Mr. President, had we not done so, we would have risked censure and scrutiny from those who oppose such legislation in Congress," the director responded.

"Director, although I know what we are doing is in

the best interest of the American public, I don't want to risk the outcry and backlash we received when revelations came out about recent NSA activities."

The FBI director jumped in, completing the thought he knew the president wished to hear. "I believe we should be fine, Mr. President."

It was another month before the neophyte junior congressman from Virginia brought the mysterious memo to the public's attention. Between the FBI director's announcement and the present-day scenario, the newly elected congressman obtained "top secret" information from an inside FBI source about the document itself. "I have an affidavit before me," he told reporters. "I shall divulge it in its entirety on the news tonight at 6:00 p.m. For now, let me say it raises serious questions as to what purpose the creation of this new agency serves in the context of our current problems."

***

In another part of the country, two senior citizens were just getting started for the morning. Nancy Lott gingerly drew back the faded green curtains from their double-wide bedroom window to get her first glimpse of the sun poking its head above the horizon. The clock read 6:01 a.m., and Lucas had gone to the bathroom to begin his morning routine. In the background, the morning news was in full swing.

"It's another gorgeous day for maritime Florida," the announcer said. "But we have just received a disturbing report of a new congressional multi-billion-dollar program that the FBI approved over a year ago. On the surface, it appears the CJIS has already initiated work on this front. However, a top-level source has informed Congressman Elijah Remington from Virginia of the ominous nature of this program that purportedly links human psychological behavior with brain wave technology at the sub-molecular level. Even as I speak, work is being carried out on how it might be possible to incorporate these studies into a practical format that will allow our government to reeducate Americans to accept programs that are beneficial and favorable to their well-being. Tune in tonight at six for a special broadcast featuring the congressman. The shocking news related to the formation of this new government department will rock America to its core."

Nancy called out from the small alcove that served as their family room and office. "Lucas, did you hear that? The government is stepping into Psychology 101. What do you think?"

Lucas stood silent as he emptied the granules from the Folger's jar into his "Greatest Dad" cup he had purchased two years ago as a joke for his first coffee of the morning. "I guess they weren't satisfied with manipulating the election," he said with a grin.

"You still haven't let that go? Trump lost and that's the end of it! No amount of wishful thinking will change that," his wife responded from her dark green Lazy-Boy next to the kitchen. "Give it up! I'm tired of hearing it."

"Think what you will," Lucas said in a shaking voice. "I still think they stole the election from him. What's this about a new government program that will examine our brains? There's not enough gray matter left in mine to worry about," Lucas said with a chuckle. "Don't these people have anything better to do with their time?"

"I don't know," replied Nancy, the weight of all her deep thinking taking its toll on her.

***

By January 1, 2020, there were over 20 million infections worldwide from SAR-COV-2, and more than 340,000 people had died from COVID-19. No one ever imagined their lives would change so drastically, or that the government we trusted would use the disease as a political football to mold our country into an arena of mindless, government-manipulated robots, incapable of making their own choices and decisions.

Forced to change against their will, Americans fought the inevitable for as long as possible. Lucas remarked he couldn't remember anything like this before. Already dealing with Parkinson's disease, Nancy struggled to remember anything from day to day.

Sandwiched between the humdrum of the daily news and the charts depicting the mounting numbers being infected by the virus were the constant reminders and updates on how to take care of ourselves during what was being described as possibly the world's worst pandemic ever. But unfortunately, the news seemed to be targeted at the new "elite," America's educated millennials, who, for the most part, walked totally in sync with the "science" of the day.

Americans were admonished to wear masks (if they could find them) and maintain a social distance of six feet from others around them. The conditioning happened right under their noses. People were too scared to do anything other than blindly obey. The media endlessly bombarded their viewers with their prejudiced perceptions, with just enough genuine information to have people believe that all of it was true. Psychology 101 at its best!

Following the mandates was easier than railing against the system. Other than financially, it didn't seem the government's dictates were hurting anyone. It was easy wearing a mask, and for the most part, social distancing was simple. People like Nancy and Lucas, who lived near Jacksonville, Florida, were fortunate: they never had to shut down completely, as others did in states like California.

Someone asked Lucas how something like this could happen, and how the people accepted it. "I don't know. The government told us what to do, and that was it. We didn't really notice the change until it was too late."

***

Malcolm X is quoted as saying, "The media's the most powerful entity on earth. They have the power to make the innocent guilty and to make the guilty innocent, and that's power. Because they control the minds of the masses."

One night, Lucas remarked to Nancy, "It seems there is nothing the president can do that would make things right. There are no amounts of precautions or measures that would satisfy the other party."

The pundits lamented the lack of preparations that had been made, and our lack of resources, to handle the increasing number of cases coming into the hospitals and clinics. It was a heyday for Saturday Night Live.

Not long after our borders were closed, every American was admonished about things like "social distancing" and wearing a mask. So peculiar, many people would think, when Americans were concerned about Middle Eastern women wearing their burqas and hiding their faces in public only a few short years ago.

Within days, at the repeated insistence of public health officials, the government imposed sanctions on businesses, forcing many of them to close. Restaurants and recreational facilities were the first to go. This was the beginning. As in pre-war Germany, some people wondered if the mandates created precedents that could not be undone. Of necessity, to lessen our burden, the government voted for "stimulus" packages worth trillions of dollars to give monetary relief to the people and businesses affected. The government printing presses were working overtime.

***

The public was scared, and the bizarre thing was that the media fed into the frenzy. People were ordered to stay out of public places and remain home except for medical emergencies and food. They were told that the government had the resources to take care of its constituents. Lucas later said this was how it all started.

It hit the elderly first and the hardest. It was

eventually discovered that the numbers had been exaggerated to engender more support for the draconian measures the government was imposing. It was strange. Hospitals received extra funds each time they certified a COVID-19 death. Lucas reflected that it would have been a great time to own hospital stocks. Even at the worst of times, profits are made from other people's misery. It is said that fear can topple governments, and avarice is the current that carries the malady. Was there a deeper element that stood to gain from our government's decisions?

Seniors were terrified to leave their houses; even later, they feared going out on their own when the numbers had been debunked. "It's tough getting old," Lucas commented.

***

Nancy and Lucas lived on the western shore of St. John's River on Fleming Island. They faced what used to be the fruit orchards on the eastern shore, now dotted by the few remaining pastel-colored million-dollar homes on the other side of the swift-moving river. The rippling waters at sunset, and highwaymen colors, were all that most men ever dreamed of. Lucas was content.

Nancy had her own problems. She had been diagnosed with Parkinson's two years ago and struggled each day just to get by. Though it would be years with the new medications, she felt more of her life slipping away each day. She was one of those persons that President Trump's Dr. Fauci referred to as having "underlying conditions." She feared going out, even to get her medications. So, she would hunker down in the basement of her fears.

How many years had it been she'd been married to Lucas? Too many to count. Lucas was not a bad person. He was just so preoccupied with himself and didn't care what happened to anyone else. He had his own routines. He thought he was the center of his wife's universe; everything revolved around him. As a result, he always had self-righteous justifications for his actions.

Their house, if you could call it that, was an old apartment nestled between two older complexes on the shore. True, their back door faced the water, but that was all they could afford. Lucas played with his town of miniature buildings and houses by day and watched sports on TV at night. His one claim to fame was that he could name every football player who had ever lived. They had three children, one who had moved to Ohio, the other to Washington State, and a third who stayed behind (she needed her parent's attention regularly). Their small unit reflected the lives of others in different parts of the country.

***

The news was more devastating each day as more and more people succumbed to the deadly disease. The virus scared everyone, and many feared leaving their homes, creating their own little prisons. It seemed no one was immune. As a result, tens of millions of people were afraid to venture outside, as if breathing the very air would be enough to transmit the disease. It turns out they were right to be frightened.

Politicians seemed content to milk the virus for all it was worth. In the shadows, the government schemed, a sleeping giant intent on one goal: subjugating its citizens to Orwell's Big Brother machinations.

Most people learned to live within the parameters of the new disease. They ordered their meals and medications on their phones or online and waited in the pickup lines outside pharmacies, grocery stores, or restaurants for their "take-out" orders. Occasionally they would have their meals delivered to their door.

Lucas didn't like paying extra for his deliveries and balked at waiting in line. "Nancy," he said on the third morning of their confinement, "do you think this will last long?"

"Lucas, I don't know. The news doesn't seem good."

There was nothing they could do. Lucas thought long and hard about their situation and how they might continue their simple lifestyle amid the new pandemic. Then, in the evening, he announced his plan to Nancy.

"Nancy, call Clarissa and see if you can get her to buy our groceries this week."

"I don't know if that's a good idea," Nancy replied, appalled her husband would even make such a request.

"She'll be all right!" Lucas snapped. "Just tell her to wear a mask!"

"Lucas, what is the matter with you? Clarissa is our daughter!" She bit her lip. She knew she had spoken too quickly. Her husband's icy stare sent chills down her spine. She knew it was time to keep quiet.

***

There seemed to be no end to the virus. When one area stopped experiencing problems, outbreaks broke out in other areas. Strangely, there were fewer outbreaks in regions like Florida that had allowed their businesses to reopen sooner. But that was something historians could investigate later.

"Lucas, I'm worried about Clarissa going into the stores all the time for us," Nancy broached the subject with her husband another day. "Can't you pay someone to go shopping for us?"

Time hung for an instant, like the hands on a Dali clock. "Lucas, you have plenty of money to buy your toys! You spent over $30 on that new rancher last week! All I was suggesting is that perhaps it might be better to let someone else, someone who is not our daughter, do our shopping for us."

"We have someone to do our shopping for us, and it doesn't cost us a cent, other than a few dollars I have to give her for gas. We don't even have to use our own car!" Lucas retorted, his eyes rolling to the back of his head. "I don't want to hear any more about this, unless you plan to do the shopping yourself!"

She felt like she'd been hit by a truck. Was Lucas this callous that he would risk his daughter's health for a few pennies?

Each person deals with life in their own way. Unfortunately, this was a pandemic, a catastrophe to bring out the best – and the worst – in everyone. Clarissa contracted the virus three weeks later. Before she was admitted to the hospital, she called her parents to tell them she had been told that she would have to self-quarantine for the next 15 days.

"Where's that leave us?" Lucas mouthed the first words that came to mind.

"Daddy, I'm sorry," she replied contritely. "I'll be out of quarantine in 14 days – tops."

Lucas hung up the phone. He had nothing more to say to her.

***

If only life had not changed. He was not a bad man, just a child who had been spoiled. If only his sister had not arrived when he was eleven, when he believed his life's direction was already assured. The young boy did not know his actions would embarrass his parents into giving him more and cause them to put him first in whatever they did. There was no way of knowing these thoughts would be ingrained into the very essence of his being. When Nancy met him nine years later, she thought his strange little traits and quirks were cute.

The pandemic caused everyone to make changes in their lives. Plexiglass barriers now separated us from the cashiers in the stores. Hand sanitizer was a staple we couldn't do without. We made sure we got no closer than the minimum six feet from anyone we encountered in the stores. The media depicted scenes of mass arrests for attending Sunday church services.

Closer to home, there was the minutiae of planning one's activities around new appointment schedules and making certain we allowed extra time for the wait lines out in our vehicles. Everyone was required to endure the "scan" and answer five COVID-19 questions before they were allowed into stores and doctors' offices. As a result, doctors saw fewer patients, and only emergency operations were permitted. It was OK – as long as we followed protocol and didn't make any waves.

Like leaves in the wind, the average person had little to say about the goings on in Congress and the White House. But, for Lucas, there were more times now that his chest pounded, and he felt the onset of migraines. His mouth dry, his temples exploding, he wondered aloud why his wife wasn't responding to his queries since his daughter had been intubated.

***

A perfect life beset with the turmoil of an event that no one could have foreseen. Actions have consequences, and sometimes those consequences create situations beyond our control. The aftermath of our government's actions had vast sweeping effects on everyone's life.

Fearful for their own lives, people did not question the subtle and not-so-subtle changes made in their lives. Governors could close businesses without reprisal and issue orders for their citizens not to gather in their places of worship, authorizations unheard of before the pandemic. State constitutions were put to the ultimate test as the authorities challenged the right of ingress and egress and chose what styles of protest to endure or reject. Finally, in the greatest test of military might ever to contain civilian unrest, tens of thousands of National Guard troops were brought into our nation's capital, purportedly to guarantee that there would be no problem with proclaiming the new president.

Understanding the psychology of the mind was the government's first step. Incorporating that knowledge into a useable, practical format capable of mass distribution was the prerogative of the scientists and engineers in the hidden labs of the CJIS in Clarksburg.

***

It seems like masks and hand sanitizers will be around forever. Social distancing signs still litter our floors. Restaurants still wipe down the tables with disinfectants after each meal. Companies like DoorDash and Grubhub frequent the airwaves advertising their home delivery services. Curtailed hours are changing in many states, and many businesses that were closed have now begun, over one year later, returning to work. Tens of thousands of businesses lost their financial battle and were forced to close their doors forever. There's word of another trillion-dollar stimulus package in the works. The government printing presses will be working overtime – again.

There are lessons to be learned from the COVID-19 scare. Perhaps the biggest lesson was that once extrapolated to its basest denominator, people are more alike than different, and that fear is a powerful tool in human psychology. We have only to look as far as the millions who willingly gave up their freedoms for the promise of a new tomorrow, free from the pandemic. Unfortunately, we still live in a world circumscribed by the fears of an imagined future contagion, secure in our belief our government will protect us, regardless of cost.

***

Congressman Elijah Remington was never seen again after he promised the special newscast. Brief mention of his failure to appear that night and a subsequent investigation into his disappearance turned up nothing, and his case was eventually relegated to the cold files. The word was that the congressman had obtained unsettling information about the formation of a new government agency that had been given the acronym M.E.E.K. That was over a year ago.

When the congressman's body was found buried in a shallow grave alongside a hiking trail in Roanoke, Virginia, fourteen months after he went missing, police immediately re-opened the case. His wallet was missing, and forensics showed that a 9mm shell had pierced his skull. However, no weapon was ever found, and there was no motive for his horrible death. Instead, rumors circulated he had been murdered, execution style.

M.E.E.K. never became public. Congress and the president were content not to raise further questions. Work went on quietly in the dark corridors of power.

***

The same day the congressman's body was found, an astute D.C. detective found a note in the congressman's pants pocket in his belongings still in his D.C. condo. The detective reported his find to his captain. Two days later, the file marked for the archive room disappeared.

Thankfully, the detective had kept a copy of the first page of the note that the congressman had planned to read the evening he was murdered. It read:

Regarding the announcement of the FBI's new M.E.E.K. department, I feel compelled to inform the public of the grave reservations I have over the mission of this new agency. Under the umbrella of the Criminal Justice Information Services (CJIS), which coincidentally is the FBI's best-funded branch, M.E.E.K. will investigate integrating human psychological research with the in-depth studies of a renowned MIT psychologist.

It has come to my attention that this program, initiated 18 months ago under the auspices of the House Internal Affairs Committee, sub-section H.R. 89417H, and referred to in the halls of Information Services as HGM, is enough to freeze the blood that runs through the veins of every American who still harbors the impression that our government is exempt from the evils of totalitarianism, mind control, indoctrination, and forced reeducation.

Americans are in for a rude awakening if HGM becomes a reality. One need only look back at the excesses of fascism in Germany, and communism in Russia and China, for a glimpse of what life is like in an Orwellian dystopia. We cannot allow HGM, the Department of "Human Genetic Modification," to continue unabated. Such a concept is anathema to all Americans.

***

In the aftermath of the COVID-19 pandemic, is it possible that our government has manipulated us into a corner from which there is no way out? Are we scapegoats of a more sinister plan for the future? Has America turned an ill-fated corner? You decide!

About the Author

Richard A. Marschall received his B.S. degree from Townson University. His postgraduate work was at Western Illinois University. His short stories have been published in anthologies from the Space Coast Writers' Guild and the Brevard Scribblers. His poems have appeared in publications by Poetry Soup and Eber & Wein. He served as Literary Editor of the Brevard Scribblers for 2018-2019.

NIH National Institutes of Health
Turning Discovery Into Health

Health Information | Grants & Funding | News & Events | Research & Training

COVID-19 COVID.gov | NIH Research information | Español | NIH staff guidance (NIH Only)

Home » News & Events » News Releases

NEWS RELEASES

Wednesday, June 15, 2022

NIAID Director Fauci Tests Positive for COVID-19

Today, Anthony S. Fauci, M.D., director of the National Institute of Allergy and Infectious Diseases (NIAID), part of the National Institutes of Health, and Chief Medical Advisor to President Biden, tested positive for COVID-19 on a rapid antigen test. He is fully vaccinated and has been boosted twice. He is currently experiencing mild symptoms. Dr. Fauci will isolate and continue to work from his home. He has not recently been in close contact with President Biden or other senior government officials. Dr. Fauci will follow the COVID-19 guidelines of the Centers for Disease Control and Prevention and medical advice from his physician and return to the NIH when he tests negative.

NIAID conducts and supports research—at NIH, throughout the United States, and worldwide — to study the causes of infectious and immune-mediated diseases, and to develop better means of preventing, diagnosing, and treating these illnesses. News releases, fact sheets and other NIAID-related materials are available on the NIAID website.

https://www.nih.gov/.../niaid-director-fauci-tests-positive-covid-19

"Follow the science" became a mantra for politicians and health officials during the pandemic. Dr. Fauci was the public face of the government's response to COVID-19 for several years. As with President Biden, Dr. Fauci is fully vaccinated and has been boosted twice.

He tested positive for COVID-19 on June 15, 2022.

In the classical sense, that's not how a vaccine works.

AFTERMATH

"There is an urgent need for global stakeholders to cooperate in simultaneously managing the direct consequences of the COVID-19 crisis. To improve the state of the world, the World Economic Forum is starting The Great Reset initiative."

https://www.weforum.org/great-reset/

World Economic Forum (WEF). June 2020.

PREDICTIONS OF LIFE AFTER COVID-19

Ashley McGrath

More than a year after the World Health Organization declared COVID-19 a pandemic, there is a light at the end of the tunnel, thanks in large part to vaccines. The coronavirus may eventually disappear like the Spanish Flu of 1918 did. But the aftermath of the pandemic will have long-term consequences.

People will eventually stop wearing masks and social distancing, but germaphobes will take hygiene more seriously than before the pandemic. They will wash their hands thoroughly and use hand sanitizer regularly. They may be less affectionate when greeting people other than their relatives and avoid those who are sick. They might take a booster shot or a third COVID-19 vaccine during the next 12 months. These precautions will be taken to prevent another pandemic.

Some individuals who contract the coronavirus will have lingering symptoms known as "long-haul COVID." Examples of symptoms are shortness of breath, coughing, and loss of smell or taste and hair, which could last months. Another health problem is organ failure, as explained on the Mayo Clinic's website. According to the American Psychological Association, there will also be mental health issues, such as post-traumatic stress disorder and brain fatigue, which will require ongoing treatment.

Many companies (particularly small businesses) and people who lost their jobs suffered significant financial losses during the pandemic. Nevertheless, there will be a gradual period of economic recovery. After reopening, restaurants, bars, hair salons, movie theaters, gyms, and other establishments will recoup some of their losses. The economy will also be stimulated by tourism, with more people traveling out of state or the country.

Most children will return to face-to-face school by the fall. There is a concern they will be behind academically, but kids have different learning styles and learn at different paces. Remote learning is a preferred educational option for some students, so I believe they will remain at grade level with hard work.

Many adults will return to their workplaces, while some will continue to work at home. Not having to dress professionally and drive to work saves time, increasing productivity. The benefits of telecommuting may entice people to work at home.

At the rate COVID-19 vaccines are being distributed, I predict people will be able to spend Thanksgiving and Christmas with their relatives this year. Regardless of what life will be like after the coronavirus is eradicated, we will remember the pandemic for years to come.

About the Author

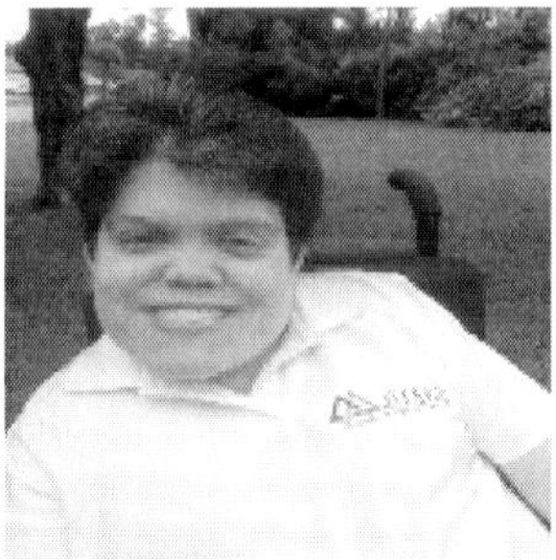

Ashley McGrath is a quality analyst for a call monitoring company. A lifelong Brevard County resident and a University of Central Florida alumna, Ashley published her autobiography *UnabASHed by Disability* in 2014. Her writing has appeared in online columns, a disability blog, local publications, and several anthologies. A Lifetime Member of the Space Coast Writers' Guild, Ashley is a Director-at-Large and the coordinator of its Don Argo Award. She is also active at her church in Palm Bay.

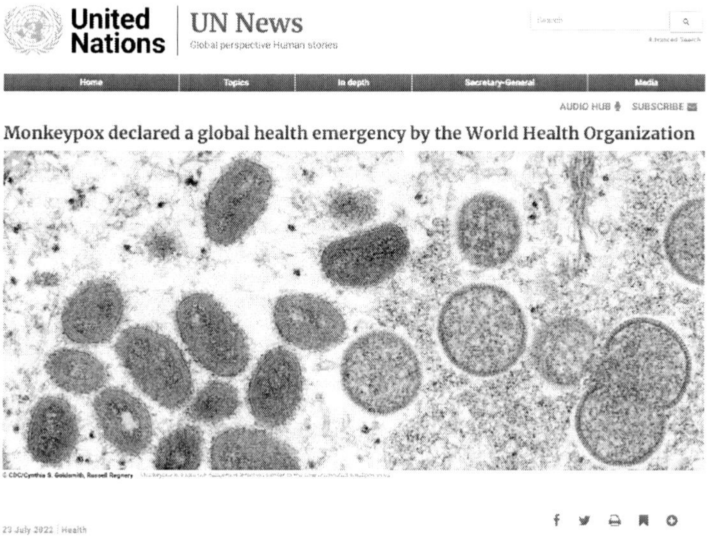

https://news.un.org/en/story/2022/07/1123152

It was only a matter of time before the next pandemic candidate arrived. In this case, it was monkeypox, which the WHO declared a "global health emergency" on July 23, 2022. Transmission mechanisms were kept highly abstract. Vaccines were in short supply. The public was scared and confused.

Did we learn anything at all from COVID-19?

AFTERMATH

"Monkeypox is an outbreak that has spread around the world rapidly, through new modes of transmission about which we understand 'too little', and which meets the criteria of an emergency under International Health Regulations.

For all of these reasons, I have decided that the global monkeypox outbreak represents a public health emergency of international concern."

WHO Director Tedros Adhanom Gebreyesus
July 23, 2022

Rehab and Relocation During COVID-19

Richard McNamara

Having been around for over 70 years (there's a grandpa-ism), I can assure you is that everyone is different. I can add that not only are they different, but they also react to the same stimuli in different ways. Because of these differences, I fear that some people take too much risk while others cower in fear. I like to think that I am somewhere in between.

In October of 2019, I suffered a stroke which caused me to lose function in my right side. After much therapy and hard work, I recovered partial use of my right arm and leg. I can now walk, and with a certain amount of difficulty, I can write and type. I was released from the rehab center a few weeks before Christmas 2019, and my wife Judy and I moved in with my youngest daughter, Christie. We had to move in with her because, due to accessibility problems, I couldn't return to our home in Grant. Living with Christie was fine, but it required a certain amount of adjustment on our part.

We were just about acclimated to our new surroundings and limitations when my wife's mother developed severe health issues and was in the hospital. So just before Christmas, Judy had to move to Tampa to care for her mother. With Judy in Tampa, Christie had to take time off work to care for me. She had to take me to therapy twice a week and take care of her house. When she returned to work, my oldest daughter Melissa took family leave and was my caregiver.

When Judy returned to Palm Bay, she brought her mother with her. Judy took care of two invalids while living in unfamiliar surroundings. Once Judy's mother had settled in, Melissa was able to go back to work. Judy had to drive me to therapy, take care of her mother, keep the house stocked and supplied, and cook for the whole house. I don't know how she did it.

***

I soon reached the point where I could take care of myself, and Judy's mother partially recovered, which removed an enormous burden from Judy. Then COVID-19 struck our country. My immediate family and I stayed safe, but I could not do anything except stay home due to the worry of infection. I spent my time dog sitting, reading, and writing. Judy still had to go to the grocery store and the pharmacy. Every time she went out, I was afraid she might get infected.

About this time, the number of positive cases of COVID-19 was on the rise when my buddy called me one Sunday morning, "Hey Mac. We're going down south and getting breakfast sandwiches. Do you and Judy want to go? They're not letting anyone sit inside. You order from the old take-out window and either take your food with you or eat in the car in the parking lot. So, we're going to eat in the car."

"That sounds great. We'll meet you there." We left Judy's mother in Christie's care and drove to Sebastian.

It was nice to get out of the house, and even though we did eat in our cars, there were times when people would walk up to say hello and want to shake hands. While it was great to see old friends again, and I thought that we were safe because we were social distancing, in the back of my mind, I was concerned.

Pretty soon, we met our friends for breakfast more frequently. Then we started going out with them on Saturday night and Sunday evening. When at the bar, we always sat at an outside table and stayed away from the crowds. It concerned me that hardly anyone was wearing a mask, and they were not social distancing. That's when the phone calls started.

"Hey Mac, I just heard that so and so is in the

hospital with COVID-19." I would get one of these calls every week or two. What was worse was that all the people passed away from the virus. I didn't go to their funerals because of my fear of contracting it. Then I decided that it wasn't worth it to put myself and Judy at risk by going to the bar. This was happening way too often to ignore. I told the wife, "I don't think we should go down south anymore until this COVID-19 thing is more under control."

"I think you're right," she said. "Every month, another one of our friends is dead or dying from this virus. It's bad enough that we must go grocery shopping or go to the pharmacy, but we're just taking a chance every time we go out. At least when you go to rehab and the gym, everyone wears a mask." Then my health care provider closed the gym and moved it to Viera. Since it was too far to drive, I called my health care provider. They said I could go to a private gym closer to Palm Bay and they would cover the cost. Unfortunately, Florida closed the gyms while I was looking for another place to work on my arm and leg.

***

It was about this time that we found out that some of my grandson's friends on his wrestling team were sick. It wasn't long until they curtailed the wrestling program while the school tried to figure out how to keep everyone safe. My grandson was devastated. He was constantly griping about not wrestling when another of his friends called. He found out that this friend's whole family (mother, father, and his brothers and sisters) had tested positive for COVID-19. Luckily, if you can call being sick lucky, only the father got ill, but he did not have to go to the hospital. These occurrences were a great wake-up call for my grandson.

Things got a little better when Judy's mother moved back to Tampa and took care of herself. We were still staying at Christie's house, but it was easier now that her mother could move back home. During this time, Judy started looking for a place for us to live. As I said before, we couldn't return to our home in Grant. We were fortunate because Judy found a house just around the corner from Christie. With her being so close, Judy didn't have to worry if something happened and she needed help in a hurry.

After we signed the papers for our new house, our friends came and helped in so many ways. Some painted, others cleaned, lined kitchen shelves, and hung curtains. We purchased new furniture, and Judy got a great deal on adjustable beds. But even with all of that, I still felt trapped by the virus. I continued to go to rehab twice a week, and sometimes I went to the grocery with Judy, but I stayed in the car as other people moved in and out of the store. Our friends continued to call to invite us to go out, but we wouldn't go with them because of our fear of the coronavirus.

***

Once we had made our new house livable, we continued searching for a gym where I could continue my rehab. Florida had reopened the gyms, but I was still apprehensive about going. Judy took me to one of the better gyms close to our house to look it over. After talking to one of the trainers, we found out that they had no one that was certified to take care of stroke victims. After reviewing what equipment was available and seeing they had cleaning supplies available for us to use, I signed myself and Judy up. We were going three days a week, and I felt like I was getting stronger. I was getting more movement in my arm, and my right leg worked much better. It was not long after that I was to get what I considered the highlight of 2020.

Christie came over to our house and said she had rented a cabin in the mountains near Maggie Valley, North Carolina. The plan was that during the week before Thanksgiving, the whole family would stay at the place for the entire week. Judy called her cousins in Charlotte, and they would join us there. It would be a mini family reunion. Everyone was very excited except me. I was worried because I would have to make the trip up and back, and up until this point, my longest car ride had been a round trip to Vero to my lawyer's office.

I was torn by the thought of a week in the mountains. I was excited about staying at the cabin and visiting the antique motorcycle museum, but I was concerned about riding eight hours or so in a car. I was afraid of what the trip may be like if I had to stop all the time. My body was constantly in a state of flux. Some days I had to go to the bathroom every thirty minutes. On other days I was subjected to spasms that shook my whole body. I related this to my daughter, who assured me it would be okay. She said, "Dad, I rented two rooms in a hotel near Charleston, South Carolina. That way, we will break the trip into two segments. It means each leg will be just over four hours. The trip should be a lot easier for you that way."

To my surprise, the first leg of the trip was pleasant – except for not seeing out of the windshield. I couldn't see because I let my wife ride in the passenger seat, which put me in the back seat, so my only view was out the side windows or trying to peer around the headrests to see through the windshield. We got to the hotel late in the afternoon, and after a good dinner and a night's sleep, I was ready to go. We ate a good breakfast at the hotel and left Charleston for Maggie Valley.

The second half of the trip was physically easier on me than the first half, but mentally it was making me crazy." Who taught you how to drive," I asked her when we hit the heavy traffic on I-26 near Ashville. "Because you're driving is making me crazy."

"You did," she hollered back as she passed two semis and an old lady doing fifty in the fast lane. "Just sit back there and be quiet like my sister and I had to do when riding in our old green van with you driving." I was doing my best not to be a back seat driver, but it wasn't working. We finally got off the interstate and my anxiety about not being in control of the vehicle subsided, so I was able to enjoy the scenery.

After arriving at the cabin, I found that I was stuck inside again most of the time. The rocky, uneven terrain around the cabin made getting from the car to the porch difficult. I couldn't make it without help. The pathway to the fire pit outback was so treacherous and steep that I wouldn't even try. So, while the rest of the family is roasting marshmallows and making s'mores, I'm stuck in the living room reading and looking out the picture windows at the mountains.

On the second day, Judy's cousins Eric and David took me to the motorcycle museum, but it was tough for me to navigate the parking lot with all the rocks and the sloping ground. Once I reached the sidewalk, I was exhausted. The sidewalk was concrete and very smooth, but it sloped up to the building. It was so steep I didn't think I would make it up the hill, so Eric ran up to the lobby and got me a wheelchair so that I could get to the museum lobby. The museum was terrific. We spent several hours viewing all the exhibits, and afterward, we went down the road to a small restaurant for a late lunch. Like the museum, it was very hard for me to get to and from the car because of the rocky, sloping driveway. On the way back to the cabin, I thanked both Eric and David for their help, but I didn't feel any better.

I spent the remaining days at the cabin reading and taking pictures from the deck while the rest of the crew did the sightseeing. They always asked if I wanted to go and even tried to lure me with promises of doing whatever it took to get me to and from the car to wherever they were going. I refused because I knew it would be difficult, or in some cases, impossible, for me to get around, and I didn't want to spend my time sitting in the car. On their return, they would show me the photos of where they had been and I heard all the great stories, but it's not the same as being there yourself.

***

All too soon, our week came to an end. We loaded all our belongings and newly acquired treasures into the car, lashed our fresh-cut Carolina Christmas trees to the roof rack, and headed for Florida. As we were once again braving the traffic on I-26 around Asheville, Christie asked, "You want me to call the same hotel we stayed at before and get us rooms for tonight?"

"You know, if you feel like driving non-stop, I think I can handle the trip."

"Okay," she said. "We'll just stop for potty breaks and junk food."

I felt a lot better once we were back in Florida. The ground was a lot flatter, and the temperature wasn't in the thirties and forties. Now, all we had to worry about was Thanksgiving and Christmas. The daughters decided we would celebrate both events at our new house, so there was a flurry of activity as soon as we got back because it was only a few days before Thanksgiving. Every day something was baking in the oven or cooking on the stove. Judy made trips to the grocery store every day, and the house was full of pies, candies, and cookies. On Thursday morning, I awoke to a house full of people putting up tables and chairs, decorating the house with wreaths and flowers, and the inevitable smell of turkey in the oven. To say it was a joyous occasion would be an understatement.

Sometime between Thanksgiving and Christmas, my health care provider said that I had no more rehab treatments available. We asked for extensions, and even with recommendations from my neurologist, they refused to give me any more treatments. Their only response was, "Call us after the new year." So, at this point, all I could do was continue going to the gym.

***

Christmas was just like Thanksgiving, with cooking, cleaning, and decorating, but I was definitely not comfortable with the whole house full of people. Without sounding like Scrooge, being around some of the people made me very nervous because they had recently tested positive for COVID-19, and frankly, even though they had been quarantined, I didn't want to be that close to them. I did my best to stay away by eating and opening gifts in my lift chair, but everyone still wanted to give hugs. Finally, after they all left, I felt better, but I had to take heat from Judy about my aloofness during dinner and the gift exchange.

When the new year came and went, I felt a little better because, after all, things were supposed to be better in 2021. I was deeply troubled by all the controversy concerning the election, and every night the news reported more and more infections and deaths. But that soon took a rear seat to my revelations concerning the gym I was going to. I was working out when I noticed that I was the only person wearing a mask other than the employees. When we got home, I told Judy, "I don't think I will go to the gym anymore."

"Why not? I think it's helping you a lot."

"I do too, but I feel like a sitting duck in there because nobody is wearing a mask, and hardly anyone is wiping down the equipment after using it. I have enough problems right now without catching COVID-19 from somebody at the gym. I just feel like if I got it, I wouldn't make it through the treatment."

Now that I was no longer going to the gym, my trips outside the house were extremely limited. Occasionally, I would go with Judy to the grocery store, but I would always wait in the car to avoid the people inside the store. I spent most of my time watching television and writing, but even these activities fed my depression. Every night the TV stations reported the number of new infections and deaths from the virus. When I went online, articles filled my browser with reports of the virus and the number of deaths. Then I got some news that caused my anxiety to go through the roof.

Judy and I were watching a movie when the phone rang. I could only hear one side of the conversation, but the few things I did hear set my hair on end. "So, who was that?" I asked.

"That was Melissa. She has taken a job for six weeks doing radiology."

"What, like a traveling job?"

"Yeah. Except that it's six weeks in a hospital in New York City!"

"New York! Is she crazy?"

"I guess it's really good money. They're putting her up in a hotel and giving her per diem for food. They also have free Uber to take her to and from the hospital," Judy said. "But the worst part is she leaves tomorrow morning. We won't even be able to see her."

We both sat in front of the TV, immersed in thought. The movie continued to play, but neither of us watched. We were both in a state of shock and concerned for our daughter's safety.

***

We are now almost through February, and things are as they had been in January. Judy kept the house running, and I continued my writing, but we both knew that something would have to change. Adding to our dismay were the phone calls from Melissa. She was depressed because of her almost constant contact with the patients so ill she knew they wouldn't make it. She was also homesick. She missed her daughter and husband terribly, but she had committed to finishing her contract. We tried our best to cheer her up during the calls, but it wasn't easy.

As we moved into March, Judy and I discussed getting the vaccine. We decided that I would go first, then she would get hers. I had decided that if I could leave the house more often, maybe my feeling of being confined would lessen. To that end, I tried to get signed up for one of the vaccines. Unfortunately, this proved to be easier said than done. Every day I spent hours trying to get an appointment. All I managed to do was get more and more aggravated.

Once a month, I would meet with people I used to work with at a beachside restaurant that allowed us to sit in a back room isolated from other guests and seated us six feet apart. At least that way, we felt safe while still talking. Judy and I were returning from the March luncheon when she got a text from our healthcare provider. "You're not going to believe this," she said. "You have an appointment next Monday to get your vaccination for COVID-19. You must text back a confirmation within the next six hours. "

"Are you kidding me? How did that happen?"

"Well, according to the text, your doctor set it up for you. Are you going to go?"

"Damn right I'm going to go. If it works as well as they say it does, at least I'll be able to go out once in a while without worrying so much."

***

It's been a couple of weeks since my first injection, and I attended my sister-in-law's birthday dinner in a crowded restaurant without worrying so much. Of course, I still wore my mask in and out of the venue, but at least I could convince myself to go out. I'm only a couple of weeks away from my second dose, but in my mind, I'm already feeling much better about meeting with my friends and going places with Judy. Now, if I could just quit watching the evening news.

About the Author

Richard McNamara's interest in literature and writing goes back to his earliest years. His parents taught him to read at an early age, and he hasn't stopped yet. He is an avid collector of first edition books (mainly Florida mystery writers) and attends several writing conferences every year. During his career in electronics, he wrote many technical articles and papers. He has used that experience to move into writing fiction.

AFTERMATH

Rido / Shutterstock.com

The COVID-19 pandemic led to prolonged lockdowns in hospitals, retirement homes, and long-term care facilities. Patients and residents found their social and physical activities severely curtailed. In the aftermath, we are witnessing the struggles of our seniors trying to make up for lost time through extensive rehab sessions and mental health counseling.

AFTERMATH

A Pew Research Center study conducted in June [2020] looked at almost 10,000 U.S. adults to understand COVID moving trends better.

D'Vera Cohn is the senior writer and editor for Pew who published the findings from the coronavirus moving study. She says she asked residents who relocated during COVID-19 to identify the driving factor behind their move.

"About a quarter (28%) told us [they chose to move] because they feared getting COVID-19 if they stayed where they were living," Cohn says. "About a fifth (20%) said they wanted to be with their family, or their college campus closed (23%). A total of 18% gave financial reasons, including job loss."

Furthermore, roughly one in five of the survey participants responded that they packed up and left because of the pandemic — or know someone who was moving during COVID, shedding light on the increased moves during this time period.

Excerpt from *Coronavirus Moving Study: People Left Big Cities, Temporary Moves Spiked in First 6 Months of COVID-19 Pandemic* by Cynthia Paez Bowman (March 11, 2022).

www.mymove.com/moving/coronavirus-moving-trends/

COVID-19 Behind the Scenes in Brevard County

Rick Neale

Note: This chapter is © FLORIDA TODAY. The story first appeared in the newspaper on May 20, 2021, under the headline "COVID behind the scenes: Hospital shares how virus tested physical, mental and emotional limits." It is reproduced here (with minor edits) with permission.

Key Points

- As of May 18, 2021, Brevard County has lost 881 people to the coronavirus
- As of May 18, 2021, a total of 2,402 residents have been hospitalized since the pandemic began
- The youngest COVID-19 patient at Holmes Regional Medical Center was a toddler

---

A whoosh of air greets visitors entering the seventh-floor COVID-19 unit at Holmes Regional Medical Center, where 20 "negative-pressure rooms" equipped to contain airborne microbes house some of Brevard

County's sickest patients.

These weakened coronavirus patients battle the dread disease the same way – alone. They are isolated from spouses, children, family, and friends, surrounded by sterile medical equipment.

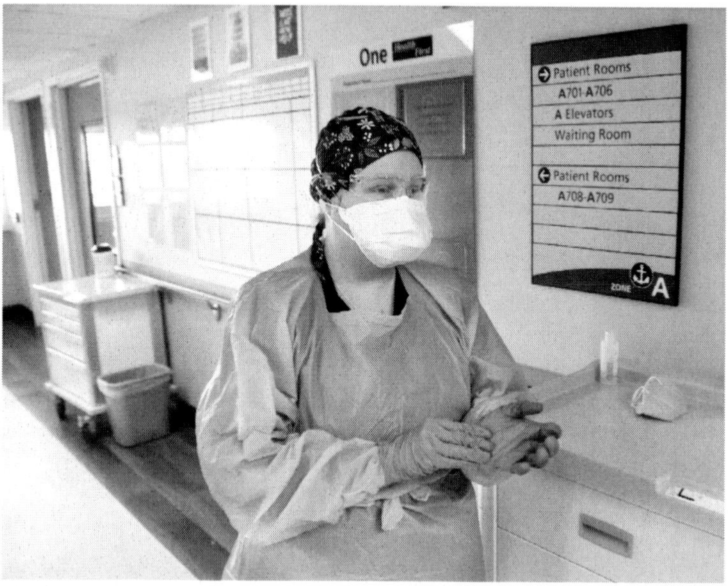

Megan Bowes, respiratory unit nursing manager at Holmes Regional Medical Center, has put on and discarded PPE more than 100 times per work shift. *Malcolm Denemark, FLORIDA TODAY.*

Some COVID-19 victims ultimately choose to forego aggressive treatment, such as intubation, and spend their final hours here on this floor, said Megan Bowes, respiratory unit nurse manager.

Three levels below – where the hospital's most debilitated coronavirus patients stay – critical care nurse Dustin Evans donned personal protective equipment and a blue hairnet before entering a coronavirus patient's room to check on her condition.

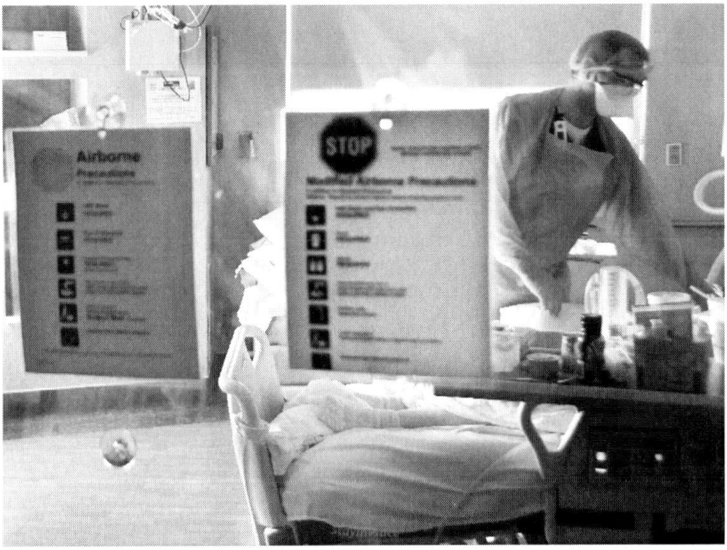

Seen through a protective glass door, Dustin Evans, a critical care nurse, attends to a COVID-19 patient in the intensive care unit on the fourth floor of Holmes Regional Medical Center in Melbourne. *Malcolm Denemark, FLORIDA TODAY.*

An older woman, the patient lay in bed amid an array of intravenous bags and tubes. Red and yellow danger signs on her room's hallway windows warned passersby of dangerous airborne contaminants.

***

A full 14 months into the COVID-19 pandemic, the Holmes medical staff continues laboring through a once-in-a-century medical crisis that has tested their physical, mental, and emotional limits.

Working stressful shifts up to 20 hours long – and fearing they might infect their families back home afterward – even veteran nurses have faced challenges unseen during their professional lives.

"Definitely very emotional. My team spends a lot of time talking about the emotions they feel. We've got the chaplain involved several times – we'll have a time out, or a moment of silence, if someone passes to honor their life," Bowes said.

"I'm used to, in my career, family being right by their side. And when the family's not there, we have a phone call to make afterward. And that phone call's so hard," she said.

"They want to know what their last words were. What their last moments were like," she said.

Bowes recalled the two sisters in their 70s or 80s from out-of-state who contracted COVID-19 while visiting Florida. Both were admitted at Holmes.

One took a turn for the worse and was transferred to the COVID-19 intensive care unit. Within a week, she decided to pass away peacefully rather than be intubated. Nurses helped her say goodbye to her sister via FaceTime.

"Which was very sad and challenging for us because that's a very emotional time. She too passed away from the virus about a week later," Bowes said.

***

FLORIDA TODAY took a rare behind-the-scenes tour of two Holmes COVID-19 units and met with eight clinical supervisors to discuss the pandemic's past, present, and future from the front lines of a world close to downtown Melbourne – but off-limits to the public.

Last year, company spokesperson Lance Skelly said Health First established an off-site coronavirus command center and spent millions of dollars building out more than 200 negative-pressure rooms to handle COVID-19 patients and "PUIers" (person under investigation for possible coronavirus).

"At the height of COVID, we had those rooms filled

up in our ICU and other ICUs – to where we had to open up another separate ICU designated for these patients," Aaron Edwards, who manages the medical ICU, said of conditions inside Holmes.

Angela Glover, cardiovascular ICU manager, has three young daughters. She said returning home from work during the pandemic's early days was "terrifying."

"We didn't know much about it. When we'd go home, you'd change clothes in your garage, go straight to the shower, didn't want anything to come into the house," Glover said.

"You didn't want to bring anything home to your family," she said.

Negative-pressure rooms were even added to the Holmes birth suites.

"We had COVID-positive moms that came in at the beginning of the pandemic," said Bridget Coggins, manager of labor birth and the neonatal ICU.

"We did not allow any visitors for COVID-positive, so they had to labor and deliver by themselves – and then be separated from their babies," Coggins said.

***

Skelly said that since January 1, Health First's hospitals in Melbourne, Palm Bay, Viera, and Cape Canaveral had admitted 992 COVID-19 patients.

"We've really had three 'waves' – each progressively bigger. One in March-April 2020, one in July 2020, then the last in January-February 2021. We've seen numbers tailing downward since February," he said.

"Now, the population typically ranges from 13 to 18 patients at Holmes and 30 to 35 patients across all four hospitals," said Nicki Anderson, vice president of nursing.

The youngest COVID-19 patient treated at Holmes was a toddler.

"There's no rhyme or reason. And a lot of patients talk about that. They're like, 'My brother-in-law got it. He's morbidly obese, and he's fine – and now I'm in here, struggling for my life,'" Bowes said.

"A lot of, 'Why me?'" she said.

Clinical supervisors at Holmes Regional Medical Center in Melbourne discuss COVID-19 and the patients they have treated during the pandemic. *Malcolm Denemark, FLORIDA TODAY.*

***

As Florida's coronavirus vaccinations increase by the day, hospitalizations and deaths are finally decreasing across the state. But the hospital's front-line medical workers worry that politicization and social-media misinformation will convince many to skip getting their shots.

"We sometimes have patients here who are really struggling – but still refuse to believe that this is COVID," Bowes said.

"I mean, up until the very end, they're like, 'Nope, sorry. This isn't COVID. I don't know what you guys are doing to me.'" Bowes said.

Last week, the Food and Drug Administration authorized the Pfizer COVID-19 vaccine for adolescents ages 12 to 15. Health First officials encourage teenagers and young adults to get vaccinated, considering upcoming summer camps, jobs, sports, Memorial Day, and the Fourth of July.

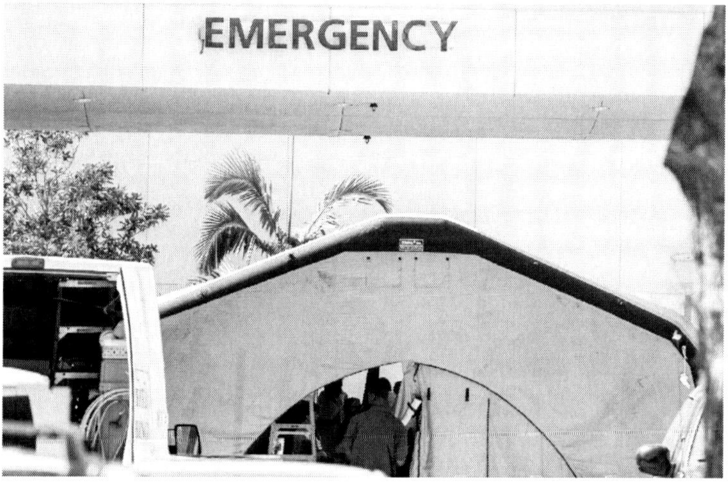

In March 2020, Health First erected a triage tent to deal with an expected influx of COVID-19 patients outside Holmes Regional Medical Center in Melbourne. *Craig Bailey, FLORIDA TODAY.*

"We typically see an uptick about a week after every major holiday," Bowes said, triggering nodding and murmurs of agreement from her fellow supervisors.

"As a nursing staff, we're prepared for it. There was a huge uptick right after the winter holidays: New

Year's, Christmas, and Easter. Then, we saw it with spring break," she said.

"We saw it after Thanksgiving last year," Anderson added. "Labor Day. Any holiday that you had a large gathering, we could always count on seven to 14 days later – suddenly, our numbers would climb," she said.

***

Last week, anti-mask demonstrators gathered in Viera outside a Brevard County School Board meeting to protest the district's mask policy. Similar school mask debates have erupted in neighboring Volusia and Seminole counties.

"I've had the great mask debate with my family members," said Nichole McKee, who manages the surgical ICU.

"And it's crazy because they'll walk around with it under their nose. You're not serving any purpose there," McKee said.

"I think that COVID hasn't directly impacted them. Or someone that they've known who had it had very mild symptoms, so it's not a big deal to them," she said.

"But they haven't been the one who sits and holds hands with a dying person. Or talked to somebody's family member on the phone to let them know that they're not going to make it out of the hospital."

***

Before entering a coronavirus patient's room, a nurse spends a minute or so donning a fitted N95 "duckbill" mask, a sleeved plastic gown belt-tied around the waist, hand sanitizer, rubber gloves, a second mask to keep the N95 clean, and protective eyewear. Finally, the nurse discards the PPE before exiting the room.

Bowes said nurses could repeat this process 100 times per shift – or more.

Last week, citing workload and dedication during the pandemic, Health First announced that employees would be granted up to 40 additional hours of paid leave, depending on job status and employment tenure.

"During all of this, none of your normal emergencies stopped or slowed down. The ER kept running. There was no stopping strokes, traumas, or cardiac issues. Patients kept coming at the normal rate," McKee said.

***

Nancy Alfred, a second-grade teacher at Roy Allen Elementary in Melbourne, nearly succumbed to the virus in December. She was admitted to the Holmes emergency room on December 8 with such dangerously low oxygen levels that she had difficulty speaking.

Nancy Alfred, a second-grade teacher at Roy Allen Elementary in Melbourne, gets emotional talking about Holmes Regional Medical Center staff during her COVID-19 hospitalization. *Malcolm Denemark, FLORIDA TODAY.*

"I was panicking. I couldn't breathe. That was the worst nightmare I've ever experienced. I'm going to be 48 in June. This is the worst feeling I've ever felt in my life," she said.

Alfred teetered at the brink of going on a respirator during her 17-day stay, but she recovered and was discharged on Christmas Eve. Still, she had to use an oxygen tank to assist her breathing for another month.

Alfred dabbed away tears with a tissue while discussing her nurses and doctors. "I felt that God was with me through all of the people's faces," she said.

***

Anderson said a host of other hospital staffers, including entry-level positions such as housekeeping, environmental services, and sterile-processing workers, have played critical roles in the battle against COVID-19.

"The front-line nurses (and certified nursing assistants), respiratory therapists, and the whole healthcare team have done the lion's share of the work. They've seen the tragedies occurring at the bedside. And yet, they are resilient," Anderson said, tearing up behind her mask.

"This is a remarkable team. And I am so extremely proud of them. Because they keep coming back, and they keep doing their best every single day, despite what's going on."

Toll of the COVID-19 Pandemic

As of Wednesday, May 19, 2021, 884 Brevard County residents had died, and 2,412 were hospitalized since the COVID-19 pandemic began in March 2020.

At least 416 Brevard deaths were residents or staff of long-term care facilities, such as nursing homes.

A total of 42,440 coronavirus cases have now been confirmed in the county. Of those, 41,701 were Florida residents and 739 were not. The median age of positive cases is 43.

The Florida Department of Health has detected 139 variants of the COVID-19 virus in Brevard.

On the vaccination front, 263,666 Brevard County residents - or 43.1% of the county population - have received at least one dose of a coronavirus vaccine. Of those, 216,340 residents, or 35.4% of the county population, have received all recommended doses.

About the Author

Rick Neale is the South Brevard Watchdog Reporter at FLORIDA TODAY, where he has worked since 2004. A former USA Today correspondent, the Melbourne resident previously reported in Ohio for the Port Clinton News Herald and Fremont News-Messenger. He helped FLORIDA TODAY win first place in breaking news in the 2019 Florida Society of News Editors contest by covering Hurricane Michael from a Panama City hotel that suffered extensive damages during the Category 5 storm. Contact Neale at (321) 242-3638 or rneale@floridatoday.com. Twitter: @RickNeale1.

AFTERMATH

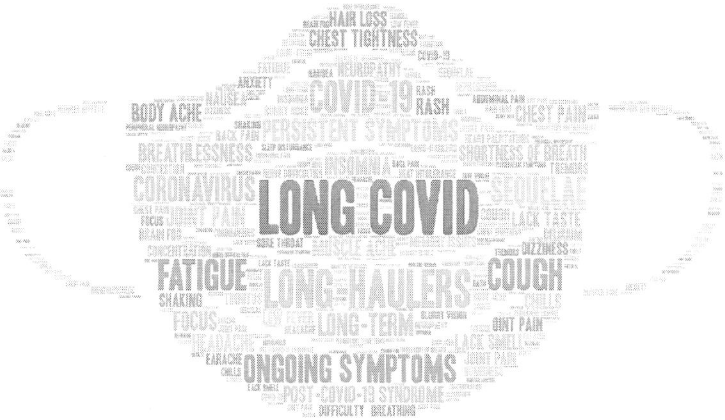

arloo / Shutterstock.com

According to the Mayo Clinic, "long COVID" involves a variety of new, returning, or ongoing symptoms that people experience more than four weeks after first getting COVID-19. In some people, long COVID lasts months or years or causes disability.

The long-term impacts on our health care systems due to a surge in long COVID cases is unknown – but likely to be very significant.

AFTERMATH

Gov. Ron DeSantis said Sunday the state remains on the right track in its response to the coronavirus outbreak despite a mounting number of new cases.

"If you follow the guidelines, you're fine," he said while speaking at Ascension Sacred Heart Hospital in Pensacola. "We have Universal Studios, they have this great safety plan, and they have the different things — [requiring a] mask, all this other stuff — [and] they've done a great job. They haven't had any problems. And it doesn't take a lot, you've just got to be careful."

His comments came as Florida reported 8,530 new coronavirus positive tests on Sunday, a day after the state set a record for a single-day increase with Saturday's 9,585 reported cases. The second highest day was Friday, with 8,942.

The state's death toll was 3,419, up 29 from the day before.

Steven Lemongello, Orlando Sentinel
June 28, 2020

LEARNING TO LIVE AFTER THE VIRUS

Stacy Strickland

Introduction

The following sentence, in some form or another, has been credited to many personalities, from physicist Niels Bohr and American writer Mark Twain to movie mogul Sam Goldwyn and baseball philosopher Yogi Berra: "It is difficult to make predictions, particularly about the future."

With that in mind, this essay attempts to predict some of the impacts or expectations in the aftermath of COVID-19 due to the SARS-COV-2 novel coronavirus that hit China with such fury in late 2019 and began to spread worldwide in 2020. The number of deaths from this disease overwhelms the imagination, and the scope of its influence on nations and their future is ongoing.

***

This global virus has kept many of us in our homes and has revised our relationship with our governments, each other, and the world. It also has promoted more sophisticated use of our technology in areas as varied as education and entertainment. It has also fostered a new appreciation for the outdoors and life's simple pleasures that we took for granted.

Europe had the Black Death in the 14th century, which caused considerable societal changes in the Middle Ages. America had the 1918 Spanish Flu that caused severe repercussions in personal lives and other countries. In both cases, there was no vaccine to protect against infection and no antibiotics to treat secondary conditions associated with them. Other than home remedies (most with a poor track record of effectiveness), control efforts after the Middle Ages were similar to those that nations currently have instituted and were limited to non-pharmaceutical interventions (disinfectants, personal hygiene, face masks, quarantine, and limitations of public gatherings).

Perhaps the most recent similar event to COVID-19 is the SARS-2 (Severe Acute Respiratory Syndrome) epidemic that struck Hong Kong in 2003. (SARS later spread to several cities, including Toronto.) SARS differed from previous epidemic infectious diseases in its rapid spread, which caught the health and hospital authorities by surprise and showed them to be ill-prepared (as with the COVID-19 outbreak in America nearly two decades later).

Inadequate epidemiology information about the disease hampered the prompt application of effective control measures (like in America). In addition, because there were no specified infectious disease hospitals (unlike in America), there were difficulties in designating hospitals for isolation and treatment of SARS patients [1].

The SARS-2 epidemic in Hong Kong affected people's health and, in its aftermath, it had social, economic, and humanitarian repercussions (as in America with COVID-19). It revealed deficiencies in the public health arena and poor communications among the health policy and management levels responsible for tracking and combating the virus (also true for America). Fundamental failings in healthcare systems became obvious: overcrowded hospitals; poorly managed long-term care homes; poor ventilation in some hospitals; lack of isolation facilities; proper medical supplies and equipment (e.g., masks, gowns, ventilators).

In addition, with healthcare staff already working under pressure, there was a noticeable effect on the mental health of intensive care nursing personnel. There was also a risk to nursing personnel from certain specialized procedures, such as cardiopulmonary activities and nasogastric interventions [1]. Hong Kong has taken steps in the aftermath to remedy these situations for future epidemics, much like America has said it will do.

On the positive side of Hong Kong's aftermath, it created a sense of unity among its sectors: government and non-government organizations, medical and nursing personnel, emergency personnel, and other front-line operators. They worked together to fight against the epidemic (much like America has done during and after the surges of COVID-19). The devotion and self-sacrifice of healthcare professionals (as also witnessed by America) in the aftermath of the epidemic's first appearance and during its spread drew appreciation and praise from all sides. It has strengthened Hong Kong's resolve to better cope with the challenges of infectious diseases in the future [1]. We hope that in the aftermath of COVID-19, America will show the same resolve.

***

As the vaccines are developed and shown to be effective in the struggle against this disease, most experts agree that even as we begin to gain some control over the virus, we will feel its impact on our future for a very long time. The virus' effect on our lives and our institutions is just beginning to be realized. There are probably even more unanticipated impacts in the aftermath that we will experience.

This essay breaks down the potential changes into eight different categories: telemedicine and pharmacy development, community, information technology, economy, the generation effect, attitude towards experts, political uprising, and changing our habits.

Telemedicine and Pharmacy Development

We are seeing a shift in the concept of healthcare delivery. Telemedicine has long been considered a cost-controlling and high-convenience system. Remote office visits will probably escalate and become more acceptable as traditional care settings are overwhelmed by pandemic side effects on people's health. There are apparent personal containment-related benefits to this. For example, staying at home for a video call keeps patients off the transit system, out of the waiting room, and away from others needing more critical care.

The pandemic has raised awareness of shortcomings in child-care assistance, access to long-term care, and how few workers have access to paid family and medical leave. It also showed how inadequate our system is in handling long-term mental health issues (e.g., more homeless and more drug addiction). This situation should create widespread political support for universal family care – a single public federal fund that all contribute to and benefit from – something that helps take care of us while we work.

Coronavirus has put a particular national spotlight on the unmet needs of the growing older population in our country, says Ai-Jen Poo, the Director of National Domestic Workers Alliance and Caring Across Generations [2]. He said tens of thousands of overstretched families and professional caregivers help this segment, but he has not seen US policy to help fully support it. It may be that increased funding for the education and training of support providers could become a reality in the aftermath of the pandemic.

It could be that the shortage of healthcare workers revealed by this pandemic results in a to-be-established public health workforce. With the Peace Corps or Teach for America as models, the workforce could offer recent high school or college graduates an opportunity to gain a few years of healthcare experience before choosing a lifetime career. In addition, the group would be able to mobilize like the National Guard during moments of a national health crisis. During calm periods, it would be able to support the efforts of the healthcare system to address the social needs of patients suffering from under-treated chronic illnesses.

The coronavirus pandemic brought out the failures of our costly, inefficient, market-based system for developing, researching, and manufacturing medicines and vaccines. Private firms will not prioritize a vaccine or countermeasures for a future public health emergency unless it is profitable. Unfortunately, we have seen that it is far too late to prevent mass disruption and poor distribution of critical supplies [3]. The reality of fragile supply chains limiting the availability of new treatments must be corrected soon.

Sterling says a more resilient and efficient pharmaceutical creation and distribution method needs to be researched and developed before the next medical emergency. I believe a combination of an ongoing partnership with pharmaceutical research firms, supported by federal government divisions like the National Science Foundation, would go a long way towards keeping drug research, peer review, and early testing as top research priorities.

Community

Americans discovered on 9/11 how vulnerable we are to calamities that we thought only happened in distant lands. In recent years, the occurrence of terrible fires in California, people freezing in Texas, and devastation from hurricanes and tornadoes in the South and Midwest, have been shocking. In addition, the 2008 financial crisis and disruption of people's lives, work, play, jobs, and even homes have brought a loss of complacency in a new way for our sense of being in the now. Deborah Tannen, professor of linguistics at

Georgetown University [4], said, "Now, the 1918 flu pandemic is a sudden specter in our lives." We know now that touching things, being with other people, and breathing the air in an enclosed space can be risky. How quickly this awareness recedes will be different for different people, but it will never vanish entirely for anyone who has lived through these last few years.

It has become second nature for many of us to resist shaking hands and follow the new guidelines for showing recognition by hand waving, elbow bumping, and so on. We hesitate to touch our faces and become Lady Macbeth in our determination to keep washing our hands. "The comfort of the presence of others may be replaced by a greater comfort with absence, especially with those we don't know intimately" [4]. More and more, we will be asking ourselves if there is any good reason why we need to do this in person. It is a bit of a paradox because it creates social distance, but we are also more connected via information technology (IT). We can communicate more often with people who are physically further away, which will make us feel safer.

There is always the hope that the concept of the "common enemy" posed by the pandemic will now be understood to not distinguish between your political or religious beliefs. It provides a singularity of purpose to help us reset and regroup, as it did for those who survived the Spanish Flu of 1918, the Blitz bombing of Britain during WW II, and 9/11 of America. Unfortunately, this is not evident so far, and only time will tell. However, past events that have affected millions of people in different countries and at different times have shown this to be the case. A study of 850 enduring inter-state conflicts between 1816 and 1992 found that 75 percent ended within ten years of a significant destabilizing shock [5]. According to Professor Coleman, these societal shocks can go both ways, to either strengthen the populace or weaken it for the future. We can hope it goes for the former.

When these pandemics end (as all pandemics do, at least in their severity), there will probably be a renewed interest in making more investments in public goods, such as health and human services. It may force more people to see how our fates are linked. Perhaps convincing the neighbor who refuses to wear a mask or stay home during a pandemic because our public schools failed to teach them science or critical thinking skills. Or the restaurant that provides me a meal where cashiers and kitchen staff are denied sick leave and make me more vulnerable to illness.

Policymakers and economists should be convinced that the economy and the social order it helps support will not be sustainable if the government or industries do not provide some sort of improved social safety net or security policies for workers who lose their jobs in recessions or depressions that go on for years from extended pandemics. They should also realize that companies and nations are losing income and people from a pandemic that negates their "business as usual" approach to operations. The coronavirus has caused immense pain and suffering around the world. It will force us to reconsider who we are and what we value in the long run, with the desire for that reconsideration to last beyond the first hint of a return to normalcy.

How will religions celebrate their faith when many of them, such as Christianity, Islam, and Judaism, rely on communal activities? All religions have dealt with the challenge of keeping faith under the adverse conditions of war, persecution, or a diaspora, but never at the same time [6]. In times of quarantine, religion can challenge your concept of what "to minister and grant fellowship' means. One can see where more contemplative practices may gain popularity among those that do not quite enjoy Zoom meetings for congregating.

Information Technology

Not everything will become virtual, but more and more of life will require IT. It will also become more ingrained as a part of the usual lifestyle. Virus screening of people via an application (as with security scans for computers) may become more ubiquitous. It will become essential to invest in infrastructure to detect future viral outbreaks. This will be particularly important if immunity to COVID-19 and its variants becomes temporary.

Companies with international supply chains are dealing with shortages and bottlenecks. As a result, many companies may re-shore some of their production. The federal government should work with US industries to ensure reliable pipelines for necessary manufacturing materials, health products, and multiple sources for our electronic support systems that are in place to buffer the country against not only a pandemic but other disasters that could occur. Global IT systems that provide confidentiality, integrity, and availability will be needed to coordinate these tasks. The growth of the IT industry will continue in the aftermath.

***

Online teaching was becoming common for all levels of education even before the pandemic. We may not be able to make up for the lost time and experience for the students of 2020 and 2021 (perhaps remedial classes added to the regular curriculum can help them), but we can certainly try to make future classrooms more productive and enjoyable. Some teachers have reported that they had previously had students who struggled to attend class because they were coping with anxiety, disabilities, or living in significant pain. Digital classrooms may help overcome those obstacles and assist students who have experienced isolation and loneliness in school with a better learning atmosphere.

Parents of a child who is doing well in a home-schooled environment might welcome the continuation of this environment in the aftermath of the pandemic. However, it will require continued access to this technology and suitable supervision – a luxury not available to all families.

Some governments, under COVID-19 urgency, have been delivering PCs and broadband to all who don't have the resources. If it is later determined that all residents are entitled to these technologies, some governments, such as the US, will likely raise taxes and spending to meet these needs. Or it could be left to market forces and devolved to the individuals and families. In this case, there will be a further separation between the have and the have-nots, resulting in individuals left behind in education and the development of computer skills for the future.

***

Another concern for the aftermath is the need for children to develop empathic and socially pertinent skills through interaction with others. The negative side-effects of isolation from real interpersonal and face-to-face interaction (e.g., depression, loss of interest in schooling, lessening of personal interaction skills) have been documented. In addition, history shows that our identity and culture depend on how we interact with the real world. A generation brought up in which communication and interaction are based on a Zoom or Skype probably does not bode well for social beings who interact within organic spaces and living, touching, breathing environments.

However, some critics have pointed out it is becoming clear that remote work is a fantastic tool but an imperfect substitute for actual human contact. To be sure, colleagues with established relationships can continue working together smoothly by chatting online. Yet bringing on new coworkers, and establishing trust and teamwork with them, can be difficult to achieve via video [7]. Moreover, remote work also leaves out all spontaneous water cooler conversations and accidental meetings that ultimately foster greater productivity and innovation from spontaneous interactions.

Once companies sort out their remote work scenarios, it will be more difficult and more expensive to deny employees the option of working from home full-time. In addition, returning to a costly dorm room on a depopulated campus will not be appealing for many college students. This will force more colleges (even smaller ones) to expand their IT capabilities - -if only to stay competitive in attracting more students.

***

The pandemic aftermath may lead to a boom in virtual reality (VR) experiences. For example, some experts would like to see a VR program that helps with the socialization and mental health of those who had to self-isolate during the pandemic. Or to assist those children experiencing lingering depression and even thoughts of suicide because of closed schools. Imagine putting on some glasses and suddenly being in a classroom or other communal setting, or even being able to engage in a positive psychology intervention [8].

There may be a new generation of augmented reality (AR) sensors, such as lidar in the iPhone12 Pro, that could land in our devices and not just cellphones. It may spur the development of AR headsets from Facebook and Apple. Just as the last three decades have been spent placing human knowledge on the web, we may take all the data and metadata built up about our physical environment and make that accessible and easy to understand through AR displays.

***

Many businesses have moved to long-term work-from-home policies. But the long-term effect of the pandemic has been causing some stress from the scale (too many people) and the duration (all the time on overused networks). The original concept of corporate networks faces outward (to connect a business center with the world). With the pandemic moving the center to a depopulated outside format, network infrastructure may have to be reconfigured in the aftermath.

An even more massive migration to cloud solutions may follow. The tools that cannot be moved into the cloud will probably be redesigned or simply eliminated because they are unstable or not secure enough for corporate work. The year 2020 saw a considerable increase in sales of laptops and servers. 2021 and after may see an increase in sales of networking equipment and the physical infrastructure needed to serve a remote organization [9].

***

In the aftermath, we will probably see new game consoles, turbocharged graphic cards, and many new games to play. It makes one wonder if we want to spend even more time interacting with diversions that emphasize separation from others or find a way to be together while still being forced apart post-pandemic. We already know that continued physical separation has a negative effect on both young (for social development) and old (for depression) in the population. These psychological impacts will doubtless continue for some time after pandemic recovery.

In the aftermath, it is hoped that developers will find some middle ground between Zoom fatigue and in-person social gaming and gathering experiences. They will need to create an offering of connection and community while minimizing the risk of infection. It may involve keeping people out of the same spaces at the same time. People's get-togethers and collaboration for play have never been forced to fit within the limitations of a pandemic. This pandemic may spur creativity in how to accomplish that. The challenge of post-locked down social distancing will be to provide entertainment and games that offer safety by distancing and delight through connections to other players and "feeling" one another's presence, yet never being in the same place at the same time [9].

Economy

It will take a while before the US and the world economy are back to normal in the immediate aftermath. At this point, the economy of the US, from an economist's view, is doing reasonably well. But while there is growing confidence that an end to the pandemic could be in sight at the end of 2021, there are warnings that viable vaccines will not halt the spread of economic damage, which will probably be felt most acutely by the poorest and most vulnerable. So again, the structure of the world economy is somewhat to blame for the uneven impact.

Global poverty is on the rise for the first time since the 1998 Asian financial crisis. COVID-19 has disproportionately affected two sectors: tourism, and micro, small, and medium-sized businesses [10]. Unlike in China, the disparities caused by the COVID-19 crisis show vaccine production and delivery will likely underscore the limited capacity of most developing and least developed countries to respond to the crisis. Most developing nations do not have the needed safety nets to support their populations. The report says vaccine deployment exposed long-entrenched inequalities in the global trading system. But it does say that the crisis could be a catalyst for new, more resilient production networks based on value chains that are shorter and more regional, sustainable, and digital. It says that much will depend on the policies adopted and the ability to coordinate, both at the international and national levels.

Mass quarantine's end (or something close to it) may unleash pent-up demand for goods and services. Sales of real estate, consumer goods, entertainment venues, and food items could result in a rapid boom in the US economy. In addition, saving rates have increased dramatically during the pandemic; cash sitting in consumer bank accounts will eventually enter the economy.

While there is no indication that the coronavirus *per se* is being transmitted through commerce, one can imagine a perfect storm in which deep recessions plus mounting geopolitical tensions from such things as tariffs limit America's access to its regular supply chains. Moreover, the lack of homegrown capacity in various product markets limits the government's response to threats. [11]. So far, that has not been the case for the American economy. To be economically prepared for the next emergency, governments may need to consider adequate redundancy in supply chains – including being resilient to trade shocks from allies.

***

After the disastrous 1918 Spanish Flu and the end of WW I (it is difficult to separate the effects), many Americans sought carefree entertainment, facilitated by the introduction of cars and radios. As a result, the economy rebounded and flourished for about ten years until irrational investment sent the US and the world into the Great Depression.

Given past behavior, when the COVID-19 pandemic

is "over," people may respond with the same search for community, relief from stress, and pursuit of economic security. We can hope that there will be a concomitant search for the preparation of an economy that can support a defense against the next pandemic – which is only a question of time, as medical experts warn us.

In the aftermath, there may be a tendency among many people to start hoarding supplies. People may believe life will require a constant stash of necessary items for long-term catastrophes, given the possibility of so many emergencies (e.g., hurricanes, tornadoes, floods, fires, and future pandemics). The shortages of toilet paper and other grocery staples may become a standard of living for many Americans. Gardening, developing home water wells and waste disposal systems, home self-repairs, home generators, and food deliveries may have become ingrained in countless households. Many probably will feel no need to return to the original method of relying on others for their needs. Businesses that support this way of living will undoubtedly do well after the pandemic.

The pandemic has caused people to confront their mortality. The death of family, friends, and loved ones may, in the aftermath, cause an upsurge in people seeking legal and financial advice and services. Estate planning, wills, trusts, financial investments, and funeral planning were usually considered long-term concerns that one can eventually address as one grows older. However, with the possibility that life may be a lot shorter, and that there may not be as much time for planning long-term events, the aftermath may spawn a flurry of work for lawyers, financial advisers, bankers, and social network support groups.

The Generation Effect

The Baby Boomers (born 1946-1964) are more at risk for health concerns related to the coronavirus. However, the older adults in the US (in most cases) are probably close to retirement. They may have less time to meet their goals, but they have social safety nets, investments, and Medicare to assist during recovery after the pandemic. Therefore, the aftermath may not impact them as much as they may think.

Generation X (born 1965-1980) is probably in midlife. They may be the best prepared to cope with isolation and social distancing because they are naturally risk-averse and motivated to stay home to take care of family and aging parents. After the pandemic, they will probably have reinforced ideas for savings and be motivated to stay home to serve as their children's role models. With the scare of the pandemic, they are more aware of saving for children's college tuition and paying the mortgage and home bills. A large portion of this group lost their jobs and was impacted by the pandemic. However, another large number were already performing their jobs remotely, and their work environment may not change. In fact, their employers may provide them the option to never return to the office. The aftermath may pose both plusses and minuses for Generation X in general.

Millennials (born 1981-1996) have been recorded in various media as a group having a decline in health status [12]. The reports see an increase in the prevalence of cardiovascular and endocrine conditions (e.g., diabetes), with a large percentage having put off medical care during COVID-19 because of cost. The reports indicate that a large part of illness is behavioral-related, impacting mental and emotional well-being. Millennials have hit the job market when there are only certain employments are on the rise, while other vocations are in decline. The economic fallout from COVID-19 has only made it worse.

Generation Z (born after 1996) face the pandemic as a defining moment for their group, from social life to career plans. Moreover, theirs is the generation that probably faces the most uncertainty in the aftermath. While the pandemic is more of a health concern for older generations, Generation Z will likely be feeling the impact of the economic downturn in the immediate aftermath. As a group, they have tended to be seen by many as more frugal and practical, driving thrift store sales, avoiding debt, and living with their parents [12].

***

What will we call people born in the aftermath of COVID-19? A possible baby boom caused by lengthy unit isolation and quarantine periods during the pandemic makes this question of labeling the next generation even more interesting.

There have been many suggestions posted to the Internet: Coronials, Pandemials, Covidials, Generation Q or quaranteenies, and Generation C. Many of the names for generations that have gone before were more about different people and organizations throwing a name out there and seeing what would stick and be promoted by the media. Only time will tell what their "tag" will be.

Attitudes Toward Experts

For many years, we in America were not often focused on topics that should have consumed our attention: occasional oil shortages, threats of nuclear war, high unemployment rates among the poor, oscillating interest rates, climate change, and terrorist attacks. The country even chose a reality TV personality (with no expertise other than talking) to run the nation's affairs. As a result, many disdained the bureaucracy and expertise that allowed our government and public services to function daily.

Professor Tom Nichols of the US Naval War College says the COVID-19 crisis aftermath could change this in two ways [13]. First, he says it has already forced people back to accepting that the concept of expertise matters. It was easy to sneer at experts until a pandemic arrived, and then people wanted to hear from medical professionals like Dr. Anthony Fauci. Second, it may return Americans to taking news warnings and science advice more seriously. Or, as Nichols opined, at least that government is a matter for people who think seriously about such issues.

Americans are becoming aware again of scientific concepts like germ theory, exponential growth, the meaning of charts and graphs, medical testing procedures, and more. Unlike tobacco use or climate change, science doubters will be able to see the coronavirus's impacts immediately. "At least for the next 35 years, I think we can expect public respect for expertise in public health and epidemics to be at least partially restored" [13].

Political Uprising

There is a possibility of a political uprising or

upheaval. Not so much as the January 6 march by ragtag, ultra-rightwing fanatics on Washington's Capitol Hill. The aftermath could inspire something more like Occupy Wall Street 2.0, but by a larger, angrier populace with all types of political beliefs. Cathy O'Neil, the founder of algorithmic auditing company ORCAA and author of the book *Weapons of Math Destruction: How Big Data Increases Inequality and Threatens Democracy,* seems to think so [14]. She believes that once the health emergency is over, everyone will see the extent to which well-connected and well-resourced communities have been taken care of. In contrast, contingent and poor or stigmatized communities will have been almost destroyed.

I suppose people could be driven to such extremes after seeing what was done in Washington, DC, on January 6, 2021. However, I do not think it will come to that. More likely, people will be driven more and more to those political organizations that promote their interests which they feel have been ignored or marginalized. Most Americans (except those on the far fringe of society) are still more likely to seek change in a more orderly and safer fashion. With the changing demographics in the USA, in the aftermath of the pandemic, the country will probably see a more liberal government elected by people who think change is best by formal methods. With a growing influence at the polls, they may believe they can better influence the future and prevent the inadequate and unfair distribution of resources before the next national emergency.

There may be more interest in voting by mail-in elections. However, after conditioning folks' desire not to stand in line, and espousing the efficiency of mail-in ballots, attendance at voting sites may drop off in the aftermath. This may depend on the ruling political environment and propaganda at the time.

Changing Our Habits

Americans are not generally comfortable with radical changes in their daily activities. You can almost notice a cottage industry that makes the dreariest possible lives sound heroic, says Virginia Heffernan, author of *The Four-Hour Workweek, The Power of Habit,* and *Atomic Habits* [15]. She mentions that we should consider another author closer to the mark when discussing a pandemic: Albert Camus. In his book, *The Plague* [16], he blames the obliteration of a fictional Algerian town by an epidemic on something else: consistency. "The truth is," Camus writes of the extremely dull port town, "everyone is bored, and devotes himself to cultivating habits." The habit-bound townspeople lack imagination. It takes them far too long to realize that death is stalking them and that it's past time to stop taking the streetcar, working for money, bowling, and going to the movies.

Heffernan suggests that it will take the dual specters of autocracy and disease to get us to listen to our common sense, our imaginations, our eccentricities, and not our programming. A braver approach to everyday existence may be crucial now so that we do not fall in line with authoritarian leaders, Kant orthodoxy, and environmentally and physiologically devastating behaviors (e.g., driving gas-guzzling cars, overeating meat, wasting electricity). Our modern plague might make us recognize that we have a short time on this earth in the aftermath. As Heffernan mentions, living peacefully and meaningfully together will take much more than just making your bed every day and following the latest hot financial investment.

Maybe the phrase "we're all in this together" will hit home with more of us and make a more lasting impression as we slowly crawl our way back towards normalcy and hopefully learn something we can take with us into the aftermath.

References

[1] Hong Kong Department of Health. *Outbreak of Severe Acute Respiratory Syndrome (SARS) at Amoy Gardens, Kowloon Bay, Hong Kong: Main Findings of the Investigation.*

April 17, 2003.

[2] Poo, A.-J. *An Opening for Stronger Family Care*. Politico News Magazine, March 2020.

[3] Sterling, S. *Government Becomes Big Pharma*. Politico News Magazine, March 2020.

[4] Tannen, D. *You're the Only One I Can Tell: Inside the Language of Women's Friendships*. Ballantine Books (reprint edition), 2018.

[5] Coleman, P. *A Decline in Polarization*. Politico News Magazine, March 2020.

[6] Sullivan, A. *Religious Worship Will Look Different*. Politico News Magazine, Politico.com, March 2020.

[7] Zakaria, F. *Ten Lessons for A Post Pandemic World*. W.W. Norton & Company (reprint edition), 2021.

[8] Bradley, E. *A Boon to Virtual Reality*. Politico News Magazine, March 2020.

[9] Pesce, M. *Peering into the COVID-19 End Game*. IEEE Spectrum, 2021.

[10] United Nations Conference on Trade and Development. *COVID-19 and Tourism: Assessing the Economic Consequences*. July 2020.

[11] Tucker, T. *Stronger Domestic Supply Chains*. Roosevelt Institute, 2020.

[12] Schaeffer, K. and Rainie, L. *Experiences with the COVID-19 Outbreak Can Vary for Americans of Different Ages.* Pew Research Center, 2020.

[13] Nichols, T. *The Death of Expertise: The Campaign Against Established Knowledge and Why It Matters.* Oxford University Press, 2017.

[14] O'Neil, C. *Expect a Political Uprising.* Politico News Magazine, March 2020.

[15] Heffernan, V. *Magic and Loss: The Internet as Art.* Simon & Schuster (reprint edition), 2017.

[16] Camus, A. *The Plague.* Random House, 1949.

About the Author

Stacy Strickland has worked in systems engineering for over 26 years. He works for L3Harris Technologies in Palm Bay, Florida, as a systems engineer in the Specialty Engineering Department. He is also an instructor in computer security at Eastern Florida State

College. He worked for Boeing Company in HW/SW systems integration on the International Space Station, at Lockheed Martin on modernizing HW/SW systems for the Eastern/Western Range at Patrick AFB, and as a system safety engineer and reliability engineer for BAe Systems. He is an INCOSE CSEP, certified Architecture Professional from the SEI, and a Certified Associate Safety Professional. He was recalled to active duty for Operation Iraqi Freedom and served in Qatar. He retired from the US Navy as a Commander. He holds an MS in computer information systems and an MBA from Embry-Riddle Aeronautical University (1990). Contact him at Stacy.Strickland@L3Harris.com.

Ivan Marc / Shutterstock.com

In the aftermath of the COVID-19 pandemic, we are witnessing "The Great Resignation." Lockdowns, working from home, and social isolation forced many people to re-evaluate their work/life balance and their overall goals. Some people opted for a fresh start, perhaps by starting their own business that could be located anywhere they want to live – as long as there's a reliable Internet connection. Saying "I Quit" to the boss and giving up on climbing the corporate ladder became quite popular. The economic impact of this mass movement of talented and empowered personnel is not yet known.

AFTERMATH

The coronavirus pandemic appears to have unleashed a tidal wave of entrepreneurial activity, breaking the United States – at least temporarily – out of a decades-long start-up slump.

Americans filed paperwork to start 4.3 million businesses last year, according to data from the Census Bureau, a 24 percent increase from the year before and by far the most in the decade and a half that the government has kept track. Applications are on a pace to be even higher this year.

The surge is a striking and unexpected turnaround after a 40-year decline in U.S. entrepreneurship. In 1980, 12 percent of employers were new businesses; by 2018, the most recent year for which data is available, that share had fallen to 8 percent.

Excerpt from *Start-Up Boom in the Pandemic Is Growing Stronger* by Ben Casselman, New York Times (Aug. 19, 2021).

https://www.nytimes.com/2021/08/19/business/startup-business-creation-pandemic.html

SCHOOLS AND BUSINESSES WILL ADAPT TO THRIVE

Dana Thomas Weber

We assembled the plane in the air. As professional educators tasked with designing hybrid elementary classrooms two weeks before the first day of school, the mandated mission was daunting. I had been aware of technology lacking in my school day for some time. However, I maintained reliance on proven systems year after year. Then, COVID-19 struck, and the 2020 schoolyear arrived with a jarring knock on my fourth-grade classroom door.

The American school system suddenly seemed antiquated. Some students would learn from home via live instruction streamed by Microsoft Teams or Zoom (neither of which I had any prior experience with), and some would learn in person from their assigned desk in the traditional classroom. In other words, a hybrid learning option would be offered to Brevard County's public-school families.

Regardless of the mode of delivery, an education

steeped in high standards remained vital. There could be no excuses, not even the fact that we were technologically ill-equipped and lacked a verified strategy. In times of adversity, foundation cracks become canyons. We had a precipice to climb.

***

Florida schools confronted the challenge generated by the COVID-19 public health scare. The system was buoyed by teachers' affinity for learning, supportive parents, cooperating students, and administration teams who stepped up whatever support they could muster on a shoestring budget and stepped back to sanction the creation of individualized teaching platforms. We shared ideas during discussions held in our empty classrooms and shared resources and knowledge online via social media, YouTube, and teacher websites. Our visions sprung to life, and, in the name of student advocacy, a grassroots endeavor emerged as we adjusted to our new reality.

Insecurities and fear of failure were banished, and, after take-off, we soared (relatively speaking). The best scenarios occurred when teamwork prevailed and a new school year was approached with patience, open minds, and a willingness to learn math, science, reading, and history in a novel manner as we watched events unfold before our eyes. Teachers learned how to effectively educate in a twenty-first century that was changing by the minute. Those who embraced innovation ushered in a new era better suited to modern practices. I recognized the true benefits of educational technology and that these modifications were overdue, regardless of the current situation. I knew that I would maintain hybrid features when classrooms returned to traditional settings. Processes had been implemented that made sense during a time of hardship. Hardships are a fact of life. Retreating to the past to embrace dated techniques defies logic.

***

Forward-thinking minds overcame the pandemic's hurdles from the classroom, Main Street, and global business empires. Small businesses strived to adapt to safety guidelines, and entrepreneurial spirits mixed with creative endeavors set some apart. Employers rooted in tradition found efficiency ratings can increase while bottom-line expenses decrease with a modern approach to doing business that lets employees skip the drive to work. Out with the draining commute and in with the home office. Around town, we saw examples of restaurants, shops, non-profits, and tradespeople meeting modified needs, and by doing so, they endured.

A global economy emerged long before COVID-19. When lockdowns struck worldwide, companies and schools built on technology rather than concrete prospered. Still, it is human nature to resist change by maintaining traditional ways. Monday through Friday, nine to five, family members leave the house in a mad dash to learn and work in outdated spaces. But work does not always end at five, and emotional and learning needs are not always met inside the school building.

Consider technology that has made it possible to fit education and work into our lives at appropriate times and convenient locations. Times and places that do not dominate the daily schedule. We can shop from our couches at three in the morning and have goods delivered to our door in record time. As it turns out, New York City is no longer the only place that does not sleep. Some time ago, the Big Apple's ways rolled away from Manhattan, into our living rooms, and worldwide.

Small businesses that survived the lockdowns will learn how to compete with Wall Street, with a leg up from the World Wide Web, as our world's new commerce emerges in vivid certainty. Formal education can be accessed anytime, anywhere, and often at moments when our mental capacity is sharper. Smartphones permit business dealings after regular hours that align with differing time zones. Businesses and schools have unselfishly delivered products meting individualized needs and, in the process, triumphed.

***

Throughout the pandemic, many of us have chosen to remain within our comfort zones and pine for things to return to normal. When forced to try remote work and learning, some found the previously untapped benefits numerous. This modern attitude possesses an advantage and will lead to our future.

While COVID-19 raged at our doorsteps, good teachers remained steadfast in their efforts to educate and protect fragile young minds despite the challenges. Top employers persisted in scrutinizing new approaches to conducting business. The processes changed practices and philosophies, and when we caught a collective breath, we saw a world that looked a little different than before. Adversity will pass. When it does, what will our society do?

Pre-pandemic, the global economy was establishing itself anywhere and everywhere, and a new way of learning was taking root. Virtual curriculums, once shunned as subpar, have earned respect. Florida Virtual School [1], a full-time, accredited virtual charter school, was launched in 1997 with a mere 77 students. By the 2017-18 school year, close to 500,000 Florida minors had completed a semester as a Florida Virtual student. Since the early days of the Internet, these numbers support about 25,000 Florida families per year opting to complete school semesters online rather than in traditional brick-and-mortar classrooms.

COVID-19 is a catalyst for escalating virtual school enrollment numbers. Traditional public and private schools are ill-prepared to deliver a quality, convenient remote learning experience that many families desire. Online bachelor courses have also become commonplace, with many Florida high school students choosing dual enrollment to complete a few college courses before graduation.

Remote learning is not the best choice for all students all the time. However, it is a valid option for many, and sometimes it is the best one. Americans, if not American schools, are increasingly choosing virtual education. After COVID-19 vaccines have been dispensed and herd immunity achieved, many families will have already committed to remote learning as a tactic to prevent interruptions in education. As a result, our schools will feel pressure to reconsider conventional means of teaching and provide quality hybrid learning approaches to satisfy the needs of students and maintain necessary enrollment numbers.

***

It is interesting to consider some absences being a thing of the past in the hybrid classroom. During the

2020-21 school year, in-person students in my hybrid classroom who felt under the weather or had been exposed to COVID-19 outside of our classroom could log in to our live instruction feed and learn from home, thus preventing falling behind over missed material. Hybrid models will provide safety by preventing germ spread and choice in modes of education. Freedom to choose is the American way, and families in the position of offering their children a remote education will exercise this right post-pandemic to prevent being caught flat-footed again.

Since the Internet's inception, has the average American family ever realized its full benefit to society? Perhaps not. From this perspective, it could be said that the past year has opened our eyes to progress. Yes, we have a long way to go to reach the full potential of the 21st century. The pandemic, sadly, shut a lot down but opened our minds to a new way of life. This points to a surge in online learning for all ages in the future and an increased demand for educational technology products.

As a result, I see progressive educators and administrators using technology far more than before the shutdowns. This will weaken rigid classroom walls and open minds to modern attitudes necessary to compete in a contemporary world. A decade ago, respected universities started to offer online courses and degrees, and the online-learning stigma dissolved as these big players entered the virtual arena. In recent months, many of those taking advantage of this learning style during quarantine were sold. Not all, but many. In the article, "2021 Trends in Online College Education," published by Study.com [2], online education has consistently increased over two decades. While only 6% of graduate students took advantage of online courses in 2008, that number had jumped to 30% by 2016.

These numbers are expected to climb once the COVID-19 dust settles. According to a study by the World Economic Forum [3], an independent Swiss organization designed to merge public and private sectors for the benefit of society, global EdTech investments reached $18.66 billion US dollars in 2019. This market was already anticipated to reach $350 billion by 2025.

Professors and teachers polled in the UK, the Middle East, Asia, and Scandinavia reported increased use of technology to facilitate learning during the pandemic. These professionals also pointed out that the unexpected benefits will lead to continued use after the quarantine. Although primary learning may be best suited for in-person instruction, it should be noted that children are growing up in a tech-based world and learning styles have adapted with the introduction of online learning outlets. The definition of differentiated learning (adjusting a lesson to meet the individual needs of a student) shifted from a pile of worksheets years ago. We may now consider the benefits of a hybrid classroom to meet the needs of all students.

***

Per my experience, observations, and research, I foresee a necessity for our society to accept and utilize remote working and learning experiences steeped in innovation. Post-pandemic, this will be true for our society to meet the needs of a global and tech-driven economy and the individual needs of students, employees, and customers who are adjusting to unfamiliar paths. The pandemic created a fresh batch of pioneering professionals. Work-from-home and remote learning trends will continue their ascent, but at a breakneck pace for the vanguard entities poised to lead.

Schools and businesses willing to rethink existing frameworks can now realign finances to target efficiency and advancement through technology while seeing decreased building maintenance costs. In addition, updated curriculums will better prepare students for sophisticated careers. The tools are available; why not use them?

This generation of students will be competing on the

world stage for career and leadership opportunities within those sophisticated companies. Students will excel by using the technology for more than video games and social media. Whether the new classroom is completely virtual or a hybrid mix, online learning is the modern definition of school in 2021 and beyond.

***

While elementary, high school, and college learning is taking a more virtual path, so will the nine-to-fivers whose tasks are primarily completed in front of a screen. A Stanford University economist coined the phrase "working-from-home economy" about a decade ago [4]. Nicholas Bloom's 2014 research on working from home is in high demand as employers rethink their professional structures and contemplate permanent remote work.

In Bloom's updated joint research project with the Atlanta Federal Reserve and the University of Chicago, it was reported that the share of days working from home is expected to increase from 5 percent (pre-COVID-19) to 20 percent (post-COVID-19) in the United States. Per the study, the typical plan is to embrace flexibility by instituting a three-day work-from-home program. Companies have seen significant productivity increases while cutting overhead costs during quarantine. This will continue by embracing work-from-home policies.

COVID-19 watered seeds of change that were planted years ago. But the seeds' sprouting exposed shortcomings when the garden tried to grow in infertile soil. We will need to learn from the patches that thrived as remote learning and working become more commonplace. Along our journey, we will naturally see consequences. An abrupt, unprepared, and involuntary shift in practices during the height of the pandemic led to mixed emotions. More recently, talk of permanent changes in routine caused those lacking technology and support at home to feel insecure, and we will swiftly find inequality to assume a new appearance.

Those best suited for career growth in this new market will climb the path to success. Conversely, employees and students not equipped with proper technological tools, workspaces, or supportive partners or parents at home will fall behind and suffer educationally, emotionally, and professionally. Many skills that were well-suited to the early 2000s will be obsolete in the 2020s as we prepare our society for technological advances in our economy and schools.

Mental health impacts will lead to an increased need for outreach programs. Quarantine was devastating to children or spouses living with abuse, and if our daily routines lead to more time spent at home, these individuals will suffer more. Even those in supportive situations may experience instability as they suddenly lack face-to-face connections that occurred naturally during their workdays. Mental health professionals will be critical components of schools and businesses if we want to reap the benefits and compete globally in a tech-based world. Teams that maintain emotional well-being will rise to the top.

***

With so many employees slated to work from home indefinitely, we can revisit Nicholas Bloom's Stanford University research to ascertain various shifts in our cities' landscape, real estate, and commerce. For example, skyscrapers, many of which are outdated and in need of renovations anyway, could become obsolete. High-density workplaces are a petri dish of germs, and in a society where infectious diseases are now a focus, a disaster waiting to happen (again).

With a spotlight on low-density workspaces, elevators and mass transit systems will serve less of our society. If the masses no longer commute to the city centers, we must consider this effect on downtown businesses that cater to this daytime market. Rethinking city-centers will be vital to repurposing spaces to meet current needs as remote working becomes permanent.

Consequences are inevitable, but advantages will be

realized, too. On the flip side, is this all a comeback recipe for small towns, rural America, and Main Streets? Will mental health improve by giving employees and students the option to work and learn in a location and manner better suited to their personal preferences? In recent history, venturing farther than the suburbs was impossible for the average college graduate tethered to an office desk anchored in the city. Under remote work and learn models, many will discover the ability to relocate to rural areas. Main Streets in forgotten small towns will have a population to service again. Their smaller, more affordable buildings could be enticing to a business owner looking to downsize to a low-density, less expensive office design.

***

The phrases "work-from-home economy" and "sensemaking" were coined by Nicholas Bloom and Kevin Kniffin, an assistant professor of business at Cornell University [5]. Work-from-home and learn-from-home models will become the new normal for American and global societies. To what degree is yet to be seen, however, this adaptation will be increasingly necessary for global leadership. Inequality will be less focused on race or gender and instead on the ability to acclimate. I imagine gender scales tipped as more mothers take advantage of opportunities allowing flexibility. Their appreciation of work-life balance will align with renewed management styles crucial for hybrid or fully remote arrangements. The New World Order is there for the taking by those with fresh perspectives and capabilities.

A quote by General Eric K. Shinseki, the Army's 334th Chief of Staff, comes to mind. When appointed in 1999, he promised to streamline processes with an emphasis on technology. His philosophies were met with resistance, and his response was, "If you don't like change, you're going to like irrelevance a lot less" [6]. Buckle up, buttercups. This advice reigned in 2021.

A post-COVID-19 America will shine a light on technology, creating a robust and efficient civilization adapted to individual needs and preferences. Hybrid approaches to business and learning will become the new norm, making inequality in the workplace look a lot different in the coming years. However, those who think outside the mid-century box will be best suited for the future, with a healthy appreciation of current and in-demand technological skillsets.

This is an exciting era, transforming by the minute. Don't blink.

References

[1] Florida Virtual School. "Florida Virtual School®

(FLVS®) District Enrollment Summary." *Flvs.net.* 2019-2020 Report. Feb. 2021. https://www.flvs.net/docs/default-source/district/flvs-district-enrollment-summary.pdf

[2] Study.com. "2021 Trends in Online College Education." Feb. 2021. https://study.com/resources/online-education-trends

[3] Li, Cathy and Lalani, Farah. "The COVID-19 pandemic has changed education forever. This is how." World Economic Forum. April 2020. https://www.weforum.org/agenda/2020/04/coronavirus-education-global-covid19-online-digital-learning/

[4] Wong, May. "Stanford research provides a snapshot of a new working-from-home economy." Stanford News. June 2020. https://news.stanford.edu/2020/06/29/snapshot-new-working-home-economy/

[5] Endresen, Janice. "COVID-19's impact on work, workers, and the workplace of the future." Cornell SC Johnson College of Business. Sept. 2020. https://business.cornell.edu/hub/2020/09/25/covid-19s-impact-work-workers-workplace-future/

[6] Eidelman, Stephanie. "If You Don't Like Change, You're Going to Like Irrelevance a Lot Less." *InsideARM.com.* Oct. 2019. https://insidearm.com/news/00045509-if-you-dont-change-youre-going-irrelevanc/

About the Author

Dana Thomas Weber is a member of the Florida Writers Association and the Space Coast Writers' Guild. Dana weaves decades of the east coast, deep south, and international travel into playful storytelling tinged with bigger ideas and inexplicable ways of the universe. A teacher and the author of two well-received novels (*The Tallest Timbers* and *A Hat for Two Heads*), she enjoys connecting with fellow readers, writers, travelers, and educators via Instagram, Facebook, and Goodreads from her home in Cocoa Beach.

https://www.flvs.net

In the aftermath of the COVID-19 pandemic, many learning and training intuitions decided to keep the hybrid model of in-person and online education they were forced to adopt at very short notice in Spring 2020. Very few chose to stay completely virtual, unlike FLVS, which has been operating this way for quite some time.

AFTERMATH

In many developed economies, the 2020 lockdowns facilitated the growth of couch commerce and increased participation with food delivery services such as Uber Eats.

As a result, the Uber ride-hailing giant's online food ordering and delivery division saw a significant boost in a variety of indicators, including an estimated 152 percent year-over-year increase in revenue.

Uber Eats generated $4.8 billion in revenue in 2020 and $8.3 billion in 2021, representing a 73 percent year-on-year increase.

Excerpt from *Uber Eats Revenue and Usage Statistics* by the Editorial Staff, Apps That Deliver (Aug. 1, 2022).

https://appsthatdeliver.com/insights/uber-eats-statistics

About the Editor

Scott Tilley is president and founder of Precious Publishing, an emeritus professor at the Florida Institute of Technology, president and founder of the Center for Technology & Society, president and co-founder of Big Data Florida, Senior Fellow at the American Security Council Foundation, past president of INCOSE Space Coast, and a Space Coast Writers' Guild Fellow. His recent books include *Technically, They* (2022), *PETS* (2021), and *PANDEMIC* (2020). He holds a Ph.D. in computer science from the University of Victoria. More information: amazon.com/author/stilley.

ANTHOLOGY ALLIANCE

Anthology Alliance

The Anthology Alliance publishes collected volumes of work edited towards a unifying theme. If you are a writing group, a non-profit organization, or a school looking to assemble and publish a collection of your members' writing, contact us! We also work with aspiring authors to edit and publish short story collections, memoirs, and essays.

Anthology Alliance is an imprint of Precious Publishing. Precious Publishing specializes in taking your writing ideas from conception to fruition. We know that your stories are precious to you, and we'll do everything we can to help see your work published.

All of our books are available online from Amazon.com, usually in both print and Kindle formats. You are the author, we are the editor and publisher, and the world's biggest bookstore is the global distributor.

http://www.PreciousPublishing.biz/AnthologyAlliance

AFTERMATH

Los Angeles Times

News Alert

Aug. 01, 2022

More people are catching coronavirus a second time, heightening long COVID risk, experts say

Emerging evidence suggests that catching the coronavirus a second time can heighten long-term health risks, a worrisome development as the circulation of increasingly contagious Omicron subvariants leads to greater numbers of Californians being reinfected.

Earlier in the pandemic, it was assumed that getting infected afforded some degree of lasting protection, for perhaps a few months. As the coronavirus mutates, though, that's no longer a given. And each individual infection carries the risk not only of acute illness but also the potential to develop long COVID.

Will it ever end?

Made in United States
North Haven, CT
05 November 2026

21793610R00169